T0402747

# Environmental Challenges and Solutions

## Volume 5

**Series editor**
Robert J. Cabin, Brevard College, Brevard, NC, USA

**Aims and Scope**

The *Environmental Challenges and Solutions* series aims to improve our understanding of the Earth's most important environmental challenges, and how we might more effectively solve or at least mitigate these challenges. Books in this series focus on environmental challenges and solutions in particular geographic regions ranging from small to large spatial scales. These books provide multidisciplinary (technical, socioeconomic, political, etc.) analyses of their environmental challenges and the effectiveness of past and present efforts to address them. They conclude by offering holistic recommendations for more effectively solving these challenges now and into the future. All books are written in a concise and readable style, making them suitable for both specialists and non-specialists starting at first year graduate level.

Proposals for the book series can be sent to the Series Editor, Robert J. Cabin, at cabinrj@brevard.edu.

More information about this series at http://www.springer.com/series/11763

R. S. Deese

# Climate Change and the Future of Democracy

R. S. Deese
Division of Social Sciences
Boston University
Boston, Massachusetts, USA

ISSN 2214-2827 ISSN 2214-2835 (electronic)
Environmental Challenges and Solutions
ISBN 978-3-319-98306-6 ISBN 978-3-319-98307-3 (eBook)
https://doi.org/10.1007/978-3-319-98307-3

Library of Congress Control Number: 2018952355



This Springer imprint is published by the registered company Springer Nature Switzerland AG
The registered company address is: Gewerbestrasse 11, 6330 Cham, Switzerland

# Preface

> *"Many forms of Government have been tried and will be tried in this world of sin and woe. No one pretends that democracy is perfect or all-wise. Indeed, it has been said that democracy is the worst form of government except all those other forms that have been tried from time to time."*
>
> Winston Churchill

**Any viable and lasting strategy for addressing the challenge of climate change will require the establishment of the rule of law on a global scale in order to be effective. For such a legal framework to be stable and legitimate, it must be democratic in its origins, thus necessitating the construction of new forms of democratic accountability beyond the parameters of the nation state. A historical survey of proposals for supranational democracy indicates that this concept has become increasingly relevant as communication and transport technologies have integrated societies across the globe, and as climate change has created environmental disruptions that no nation can face alone. In this century, it may be possible to create democratically accountable global institutions that could address the challenge of climate change much more effectively than treaty arrangements among sovereign nation states. The most plausible first step in building such institutions would be to foster greater political integration among those states that are already democratic.**

In order to survive and prosper in the epoch of climate change, we will have to build a new kind of democracy that reaches far beyond the boundaries of the nation state. Climate change, primarily resulting from human activities such as deforestation and the emission of greenhouse gases, is a problem that will require unprecedented cooperation to address. Democracy is the system of government that has the best record for inspiring and maintaining human cooperation in the face of unanticipated problems and for extended periods of time. Starting from these premises, it is possible to frame a few hypotheses about how democracy and climate change will interact in the twenty-first century and beyond. First, it is likely that the disruptions caused by climate change will threaten the survival of democracy by exacerbating international and transnational conflicts. Second, it is probable that in order to survive in the age of climate change, democracy will have to grow beyond the boundaries of the nation state. And third, there is reason to expect that transnational democracy will prove to be the most effective form of governance for dealing with the global challenge of climate change.

This book will explore the work of a diverse array of scientists, intellectuals, poets, and political leaders who have advanced the idea of supranational democracy in the past and survey the various ideas that they have presented for reaching that goal. In its final analysis, this book advances the conclusion that the wisest and most principled plan for extending the reach of democracy is to form a political federation of existing democratic governments, a concept championed by Clarence Streit and Jean Monnet in the mid-twentieth century, and later refined by James R. Huntley as the global union of democracies that he called "pax democratica" (2001). Other intellectuals in the field of international relations have put their own spin on this idea in the twenty-first century. In 2006, for example, G. John Ikenberry and Anne-Marie Slaughter called for the creation of a "Concert of Democracies." Such an organization, they argued, would combine the power and creativity of the world's democratic societies to address a variety of global challenges, including "potential security consequences of climate change, from natural disasters to a fierce scramble for territory" (Ikenberry and Slaughter 2006, p. 11).

During the twentieth century, no individual advocated a global union of democracies with greater clarity, force, and persistence than Clarence Streit. A journalist who had been born in Missouri and come of age in Missoula, Montana, Streit served in the World War One and was part of the team assisting the American delegation at Versailles. After completing his education as a Rhodes scholar at Oxford, he covered the League of Nations in Geneva for the *New York Times*, and soon became convinced that a full-fledged union of democracies would be better at keeping the peace than the League could ever be. In 1938, Streit correctly perceived that Hitler would not be placated by the concessions made by Britain and France at Munich, and he raced to complete his manifesto for a union of the world's democracies. His 1939 bestseller *Union Now* advocated the immediate creation of a federal union that would include the United States, Britain and the nations of the British Commonwealth, France, Belgium, the Netherlands, Switzerland, Denmark, Norway, Sweden, and Finland. Streit reasoned that these democracies could coordinate their military assets to deter further aggression by the Axis powers and thus prevent another world war. While this plan did not come close to fruition, Streit's vision inspired enthusiasm among elites on both sides of the Atlantic, and his ideas influenced the evolution of NATO during the first decades of the Cold War (Baratta, vol. 1. 2004, p. 53–56).

In contrast to Streit, Jean Monnet's influence was not the product of grand declarations or blueprints for a new world order. Rather, his achievements emerged from his quiet and indefatigable effort to build bridges that would last between the Western democracies. He began his career as a cognac merchant, and his work in transatlantic trade, particularly with the Hudson Bay Company, gave him an early familiarity with the business and political cultures of both Britain and North America which proved indispensable to France after the beginning of World War One (Duchene 1994, p. 31–35). Applying his formidable skills in both negotiation and administration, Monnet coordinated economic and political cooperation among the Allies in both world wars. In the closing months of World War One, Monnet wrote to US President Woodrow Wilson and to French Prime Minister Georges Clemenceau to make the case that, "It is urgently necessary that the Allied democracies establish an economic union that will form the nucleus of an Economic Union of Free Peoples" (Monnet 1978, p. 79). Although such a democratic federation did not emerge from the aftermath of Versailles, Monnet remain committed to deepening economic integration and transatlantic cooperation in order to defend and advance democracy. Before the United States entered World War Two at the end of 1941, Monnet helped the Roosevelt administration generate public support for economic and military aid to antifascist forces in Europe, and was widely credited with coining the phrase "arsenal of democracy" employed by President Roosevelt to make the case for American rearmament a year before Pearl Harbor (Duchene 1994, p. 89). During the war, Churchill issued a British passport to Monnet to help him in his essential work, first in Washington, D. C., and later in Algiers. John Maynard Keynes later reflected that Monnet's work in coordinating the war effort was so effective that it "shortened the war by a year" (Duchene 1994, p. 93). In 1963, President Kennedy wrote to Monnet that, "under your inspiration, Europe has moved closer to unity in less than twenty years than it has done before in a thousand." In stunning contrast to the "emperors, kings, and dictators" who had all failed "to impose unity on Europe by force," Kennedy observed that Monnet had succeeded "in transforming Europe by the power of a constructive idea" (Garten 2016, p. 231).

In the aftermath of the World War Two, Monnet fostered the creation of the European Coal and Steel Community in 1952, laying the cornerstone of what would ultimately become the European Union (EU) (Duchene 1994, p. 225). Attending Monnet's funeral in 1979, the American diplomat George Ball took note of the music that Jean's widow, Sylvia Monnet, had selected for the service, which "consisted of songs and instrumental pieces from each member state of the European Community. Then, unexpectedly, sandwiched among the European classics, came a loud and lively rendition of 'The Battle Hymn of the Republic'" which Sylvia identified "as one of Jean's favorites" (Duchene 1994, p. 9). When we consider that Monnet's faith in the power of democratic federalism had been tempered by two world wars and the Cold War, his appreciation for Julia Ward Howe's anthem of the Union should not be surprising.

Throughout their careers, both Streit and Monnet argued that the development of strong federal ties among democratic governments was essential to maintaining peace, promoting prosperity, and protecting human rights. Monnet's efforts to advance democratic federalism laid the foundation for the European Economic Community, and later the EU. Streit's more ambitious vision of creating a full federal union between the democracies of Europe and North America did not come to pass, but it did help to inspire such enduring multilateral achievements as the North Atlantic Treaty Organization (NATO) during the early days of the Cold War

(Rosenboim 2017, p. 11). Together, institutions such as the EU and NATO, for all of their flaws and difficulties, represent the economic, strategic, and political nucleus of a potential federation of democracies that spans an ocean and two continents.

Unfortunately, in the early twentieth-first century, faith in democratic federalism, and in democracy itself, has come under sustained attack across the world. As political scholars Roberto Stefan Foa and Yascha Mounk have documented, numerous opinion polls have revealed that faith in the viability of democracy is on the decline, especially among young people (2017, p. 5). Mounk and Foa report that, by the second decade of this century, "parties and candidates that blame an allegedly corrupt political establishment for most problems, seek to concentrate power in the executive, and challenge key norms of democratic politics have achieved unprecedented successes in a large number of liberal democracies across the globe" leading to electoral victories for such demagogic figures as Donald J. Trump in the United States, Viktor Orbán in Hungary, and Rodrigo Duterte in the Philippines (2017, p. 8). In his book *The People vs. Democracy*, Mounk has charted the rise of "illiberal democracy" a form of "democracy without rights" in which charismatic leaders attack the free press, the independent judiciary, and religious or ethnic minorities in the name of a segment of the population that they define as "the people." Although Mounk sees the rise of "illiberal democracy" as emerging partially in response to the "undemocratic liberalism" of moneyed and technocratic elites, he warns that such regimes, even if they address some legitimate grievances and attain power at first through legitimate elections, will not remain democratic for long. In other words, an "illiberal democracy" will soon become an ironclad autocracy, once it has obliterated the norms and institutions that are necessary for any democratic system to function (Mounk 2018, p. 14–18).

Once a society has lost the ability to hold its leaders accountable through such institutions as a free press, an independent judiciary, and competitive elections, it is liable to be stuck with those leaders for a very long time. For Karl Popper, the chief virtue of democracy was that it afforded the public the opportunity to dismiss corrupt of incompetent leaders without bloodshed. As the political philosopher John Mueller has put it, democracy amounts to a tacit understanding between the government and governed, in the which such practices as the rights to petition, to protest, and to vote elected leaders out of office are essential for keeping the peace. In sum, "the people effectively agree not to use violence to replace the leadership, and the leadership leaves them free to dislodge it by any other means" (Mueller 2001, p. 247). Concurring with John Mueller's analysis, Steven Pinker argues that this instrument for periodic peaceful revolutions is one of the greatest achievements of the Enlightenment and is founded on the ethos that our "freedom to complain rests on an assurance that the government won't punish or silence the complainer" (Pinker 2018, p. 206). By resorting to the atavistic rhetoric of nationalism and stoking violence against both ethnic minorities and political opponents, the current generation of "populist" leaders threaten this "freedom to complain" as they set about dismantling the foundations of democracy.

The current trends of resurgent nationalism and authoritarian leadership have been gathering force since the first years of the twenty-first century, fueled by the fear of terrorism, mass migrations, and the economic dislocations engendered by globalization. In democracies across the world there is a growing sense that elected leaders have diminished power or desire to address the concerns of voters, having ceded authority to less accountable entities such as "bureaucrats . . . central banks . . . and international treaties and organizations" (Mounk 2018, p. 59). Though such essential democratic institutions as competitive elections, a free press, and an independent judiciary have extended their reach since World War Two to many nations that had never enjoyed their benefits in the past, these same institutions have become more precarious in older democracies, including the United States. The sense that globalization has heightened economic inequality has probably done the most to foment support for authoritarian movements around the world, but in recent decades the disruptive power of climate change has also emerged as a force with serious political consequences. As extreme weather events, droughts, and a scarcity of fresh water impact populations across the world, the appeal of nationalism is likely to grow, and the threats to the norms and institutions that are essential to any viable democracy are likely to multiply.

Like the man who is shocked one morning to discover that his favorite pants no longer fit, we have all experienced how quantitative change, which happens gradually, gives way to qualitative change—which seems to happen all at once. As a citizen of the United States of America for a little over half a century, I have witnessed how the practice of democracy and climate change have become incrementally, and then inextricably entwined. As I look back on the five and a half decades since my own birth, I can see that our impact on the climate has gone from being close to invisible to one of the most potent political issues of our time. A few years before I was born, the poet Robinson Jeffers wrote the following lines about the future of the earth and our cities upon it:

> *The polar ice caps are melting; the mountain glaciers*
> *Drip into rivers; all feed the ocean;*
> *Tides ebb and flow, but every year a little bit higher*
> *They will drown New York, they will drown London*
> (Jeffers [1963] 1991, p. 476)

From his perch on the rocky cliffs of northern California, Jeffers was literally and figuratively a voice crying in the wilderness. A hermitlike and bluntly misanthropic poet, he had ceased to command a popular audience in the decades of general optimism and booming economic growth that followed World War Two (Karman 2015, p. 2–4). The question of what human activity was going to the climate our planet had been raised by a few scientists such as Roger Revelle, but the question of what to do about it was not on the agenda of any politician (Weart 2008, p. 29). In a decade overshadowed by various Cold War crises and a burgeoning youth culture, it seemed for all intents and purposes that democracy and the possibility of climate change had nothing to do with each other. After the events of the past fifty years, however, it has become clear that the future of democracy on earth will be determined by how we respond to the reality of climate change.

Kenneth Pomeranz and Steven Topik have underlined the link between fossil fuels and political instability since petroleum emerged as the dominant fuel in the global economy in the mid-1960s with the inspired phrase "Running on Oil, Building on Sand" (2006, p. 252). During this period, which has been roughly congruent with my own lifetime to date, the earth has grown hotter, smaller, and more politically volatile. When I was born in 1964, the carbon concentration in the earth's atmosphere was considerably lower than it is now, at about 320 parts per million. In one sense, this was a great year for democracy, as President Lyndon Johnson signed the Civil Rights Act, forbidding discrimination on the basis of race across the United States. In another sense, this was a terrible year for democracy because in early August, Congress passed the Gulf of Tonkin Resolution, which fully authorized the American War in Vietnam. This war would last over a decade, and take the lives of approximately two million Vietnamese and over fifty-eight thousand Americans. As the U.S. Army veteran and historian Andrew Bacevich had observed, the deployment of US combat troops to Vietnam during the sixties and seventies was a "tipping point" that profoundly altered not only American foreign policy but the domestic politics of the United States as well (2008, p. 29). In addition to its considerable human cost and serious economic impact, this conflict would create political and cultural fissures in American society that still endure today, and inflict lasting damage on the credibility of the US government.

When I turned 5 years old in the summer of 1969, the concentration of carbon in the earth's atmosphere had increased to 326 parts per million. The Apollo program landed astronauts on the moon that summer and provided the first color images of the earth from space. Though Apollo had been driven by a nationalistic "space race" between the United States and the Soviet Union, these images transcended nationalism and galvanized a new global environmental consciousness. The image of the earth, which Apollo astronaut Jim Lovell had described on Christmas Eve of 1968 as "a grand oasis in the big vastness of space," was both beautiful and humbling. The first man to walk on the moon, Neil Armstrong, recalled that the earth appeared so small from that vantage point that he could block it out entirely with his thumb. When asked if this made him feel big, Armstrong responded, "No. It made me feel really, really small" (Poole 2008, p. 190).

By the time of my tenth birthday, in the momentous summer of 1974, the concentration of carbon in the atmosphere had risen to about 332 parts per million. Across southern California, air pollution from automobiles was such a severe problem that local school districts adopted a flag system to signal air quality. On the days when a small triangular red flag was hoisted on the flagpole, all physical education classes would be canceled and outdoor play discouraged, while the local San Gabriel Mountains would be obscured by a pinkish brown smog. Since the the 1970s, California has made great strides in addressing its smog problem, although it has made much less progress in addressing the problem of greenhouse gas emissions. In addition to these local concerns about air quality, there was a growing sense among climatologists that the earth was "entering an era in which man's effects on the climate will become dominant" even though some scientists still debated whether aerosol pollutants, which tend to cool the atmosphere, or greenhouse gases, which have the opposite effect, would be the more decisive factor (Weart 2008, p. 87). The population of the earth, which had been a little over three billion on the day of my birth, was now over four billion. The Hollywood film *Soylent Green*, which had been released in 1973, centered on mounting fears about population growth, but it also depicted a world severely impacted by rising temperatures due to a runaway greenhouse effect (Peterson et al. 2008). The world in the mid-1970s was also better armed. In the year that I was born, only the United States, the Soviet Union, Britain, and France possessed nuclear weapons. By the mid-1970s, China, India, Israel, and South Africa had also joined the "nuclear club," though only China and India had publicly disclosed their nuclear weapons programs. The most notable political event in the United States in 1974 was the Watergate scandal, which culminated in the resignation of Richard Nixon that August, ten years to the month after the United States had begun its war in Vietnam. In one sense, Nixon's scandal and resignation signaled the resilience of democratic institutions in the United States, especially considering it had been the free press and constitutional checks on executive power that had brought him down. In the longer term, however, the Watergate scandal created a habitual cynicism among Americans about their national institutions, and the fact that Nixon never faced criminal prosecution, and even returned to public life with the *gravitas* of an elder statesman, emboldened his successors to emulate his violation of the U.S. Constitution with a diminished fear of the consequences.

In the summer of 1979, when I turned 15, the carbon concentration in the atmosphere had risen to 339 parts per million. The concept of climate change had found its way into popular music with the Peter Gabriel song "Here Comes the Flood" which featured visions of a watery apocalypse: *If again the seas are silent, and any still survive / It'll be those who gave their island to survive* (Bowman 2016, p. 70). Catastrophic themes were gaining traction in popular culture at that time, often casting visions of climate change into the mix with a variety of fears, as in this passage from the 1979 hit "London Calling" by the Clash: *The ice age is coming, the sun's zoomin' in / Engines stop running, the wheat is growin' thin / A nuclear error, but I have no fear / 'Cause London is drowning, I, I live by the river.* Critics who deny that climate change poses a serious threat often point to the talk of a returning ice age in the 1970s as evidence that those who sound warnings about rising temperatures have changed their story and are therefore not to be believed. This oversimplifies how the debate on climate unfolded in the 1970s. While calculating the competing influence of aerosol pollutants and greenhouse gases on the earth's atmosphere, climatologists still disagreed about whether the earth would grow cooler or hotter, but there was an emerging consensus that, as one reporter for *Time* magazine put it, "The world's long streak of exceptionally good climate has probably come to an end – meaning that mankind will find it harder to grow food" (Weart 2008, p. 87). Among peer-reviewed articles on climate change in the 1970s, the majority discerned the trend toward rising global temperatures (Peterson et al. 2008).

In spite of such ubiquitous talk about the apocalypse, however, the more significant story in 1979 was probably the resurgence of *laissez-faire* economic policies around the world that has come to be known as neoliberalism. The most momentous shift toward such policies was probably the ascendancy of Deng Xiaoping in China. Among Western nations, the most dramatic shift toward neoliberalism in the 1970s took place in Britain with the rise of Margaret Thatcher to the position of

Prime Minister. Her espousal of the economic and political views of Friedrich Hayek in the United Kingdom would soon pave the way for a wave of tax cuts, deregulation, and privatization that accelerated in the United States under Ronald Reagan in the 1980s. As the historians of science Naomi Oreskes and Erik M. Conway have observed, it has become a central tenet of neoliberalism, that "capitalism and freedom go hand in hand—there can be no capitalism without freedom and no freedom without capitalism" (2011, p. 64–65). The latter part of this premise, that there can be "no freedom without capitalism," has created a political climate, especially in the United States, where any form of environmental regulation is cast as an insidious expansion of state power, and another step on what Hayek called "the road to serfdom." This trend, which began with the Reagan and Thatcher revolutions, has impeded the efforts of many governments to address the challenge of climate change.

By the year 1984, when I turned 20, the concentration of carbon in the atmosphere was now 346 parts per million. Given the high tension between the United States and the Soviet Union that year, the greatest environmental catastrophe on the minds of many was not global warming but rather the possibility of nuclear winter. In politics, the shift toward neoliberalism that had begun in the late 1970s was now an established fact. Reagan won a landslide reelection in 1984, while Thatcher, Nakasone, and Kohl pursued similar policies in Britain, Japan, and West Germany. The 1984 Summer Olympics in Los Angeles became a showcase for neoliberal policies through their overwhelming reliance on corporate sponsorship instead of public investment (Schulman 2001, p. 240). Market reforms continued to accelerate in China and would soon be attempted, though with much less success, in the Soviet Union.

In the summer of 1989, when I turned 25, the carbon concentration in the earth's atmosphere reached 355 parts per million, breaching the limit of 350 parts per million that many climatologists see as necessary for maintaining a stable climate (McKibben 1989, p. 5). This was also a year when public awareness of climate change began to rise, largely due to the efforts of NASA climatologist James Hansen and *New Yorker* writer Bill McKibben. In the realm of politics, 1989 would become a legend in its own time. Before the year was over, it had already been christened "the end of history" by an enterprising academic at the American Enterprise Institute by the name of Francis Fukuyama. From his Hegelian perspective, Fukuyama saw the fall of the Berlin Wall and the erosion of Soviet power across Eastern Europe as stark evidence that, if history is defined as the search for the best form of human government, that search was now at an end. A lasting and happy marriage of free market capitalism and liberal democracy was now the destiny of the human race, even if much of the world had not yet reached it. Fukuyama's erudite optimism and daring presentation sparked an entire cottage industry of responses by journalists and academics, but the debate that ensued tended to ignore another important event that year.

In the summer of 1989, a new and audaciously brutal form of authoritarian capitalism would reveal itself to the world, and its fortunes in the decade since have been steadily on the rise. By electing to crush the pro-democracy movement by deploying tanks and other battlefield weapons against thousands of unarmed protesters in June of 1989, Deng Xiaoping and his ideological allies in the Chinese Communist Party obliterated the neoliberal assumption that economic liberalization must lead to political liberalization. In contrast to the process of democratization that was beginning to take place in capitals such as Gdansk, Prague, and even Moscow in 1989, the brutal crackdown on the democracy movement in Beijing and throughout China showed that the Chinese government was determined to retain its Leninist one-party autocracy, even as it forged ahead with free market reforms. The *New York Times* columnist Nicholas Kristof would soon dub this hybrid form of government "Market-Leninism." Noting the success of this strategy, he observed in 1993 that, "The plan is to jettison Communism – but not Communist Party rule – and move China's nearly 1.2 billion people into . . . free-market authoritarianism" (Kristof 1993). In the early 2000s, the British historian Timothy Garton Ash described the new ideology explicitly as "authoritarian capitalism" and identified it as the "biggest potential ideological competitor to liberal democratic capitalism since the end of communism" (Ash 2008). The power of authoritarian capitalism has grown steadily in this century, and the stability of the largest authoritarian capitalist regimes has been bolstered by fossil fuel reserves, with oil and gas exports providing leverage to the Russian Federation under Putin, and a steady supply of coal fueling the rise of manufacturing, steel, and military might in China since 1989. Since solidifying the power of his regime in the first decade of this century, Vladimir Putin has refined the playbook of authoritarian capitalism and now exports it as a model for countries such as Poland, Hungary, and (since the surprise election of Donald J. Trump in 2016) the United States. The flow of political and economic refugees from the Middle East and Latin America, exacerbated to a growing degree by climate change, has heightened the appeal of this xenophobic and antidemocratic model of governance in both Europe and North America. Authoritarian capitalism, which first showed its potential for large-scale brutality in June of 1989, has proven its ability to erode democratic institutions all over the world.

By the summer of my 30th birthday in 1994, carbon concentration in the atmosphere exceeded 360 parts per million, and broad consensus was emerging that carbon emissions were altering the earth's climate. In fiction and film, a new genre of speculative fiction known as "Cli-Fi" emerged which explored the darker possibilities of climate change. For example, the movie star Kevin Costner was producing and starring in a big budget motion picture about a future earth in which the polar ice caps had completely melted. Beset by numerous production problems, *Waterworld* would not be released until the following year, to a disappointing commercial and critical reception. In light of higher temperatures, more extreme weather events, and other mounting evidence of climate change, the international community was making its first attempts to deal with this problem, under the auspices of the UN Framework Convention on Climate Change (UNFCCC). Although President George H.W. Bush had eschewed involvement with the UNFCCC, the administration of Bill Clinton promised to be more cooperative. Vice President Al Gore had made his name in the Senate for his outspoken concern on the issue of climate change, and he would have a direct hand in the negotiation of the Kyoto Protocol in

1997. However, US politics took a sharp turn to the right in 1994, when a pugnacious Representative from Georgia named Newt Gingrich led a successful Republican effort to take control of Congress. This power shift in Congress marked another milesone in the rise of neoliberalism. For more than two decades to come, a commitment to laissez-faire economics and passionate hostility to government regulation would dominate congressional politics in the United States and effectively doom federal efforts to address the problem of climate change.

By the summer of 1999, when I marked my 35th birthday, the carbon concentration in the earth's atmosphere reached 370 parts per million, an increase of well over 15%. The concept of climate change had seeped further into the culture at large, as evidenced by the lyrics of such hit songs as "All Star" by the pop rock group Smash Mouth. Reaching number two on the Billboard modern rock charts in 1999, the song reflected lyricist Greg Camp's meditations on climate change: *The ice we skate is getting pretty thin / The water's getting warm so you might as well swim / My world's on fire. How about yours?* (Emerson 2017). Increasing temperatures were becoming impossible to ignore in the 1990s, with 1999 ranking not only as the last year of the millennium, but also its hottest. In spite of this, the world's most powerful democracy was not in a position to deal effectively with climate change, distracted as it was by partisanship, a presidential sex scandal and impeachment trial, and the emerging challenge of terrorism against US embassies and military housing facilities in such places as Saudi Arabia, Tanzania, and Kenya.

By the summer of 2004, when I turned 40, the carbon concentration in the earth's atmosphere had reached 379 parts per million. By this time, I had become a father. Surveying the news about collapsing ice shelves, hurricanes, and heat waves, I had reason to be concerned about the planet that my sons would be living on by the time they were my age. Furthermore, growing signs of dysfunction in the American political system threatened to thwart international cooperation to cope with climate change. Although Al Gore, an experienced leader with a long record of commitment to the issue of climate change, had won the popular vote in 2000, he did not become president. The election of 2000 had featured a disputed vote count and widespread voter suppression in Florida and culminated in an unprecedented Supreme Court decision that placed George W. Bush in the White House. Reflecting its extensive ties to the fossil fuel industry, the second Bush administration had withdrawn the United States from the Kyoto accords, without offering any replacement. In a further blow to the reputation and stability of democratic institutions in America, the terrorist attacks of September 11, 2001, had thrust the country into an open-ended "War on Terror" that had more than a little in common with the nebulous and permanent state of war described in Orwell's *Nineteen Eighty-Four* and considerably weakened the commitment of the American public to longstanding prohibitions against arbitrary arrest, indefinite detainment, and torture (Ricks 2017, p. 257). Democratic values, which had been expanding in many parts of the world since the end of the Cold War, were now under increasing threat in the oldest and most powerful constitutional democracy on earth.

When I turned 45 in 2009, the carbon concentration in the earth's atmosphere had reached 389 parts per million, and extreme weather events had become a recurrent force to be reckoned with in American politics. The poorly managed response to Hurricane Katrina during the summer of 2005, which had left more than a thousand dead in New Orleans and the surrounding region, inflicted irreparable damage on the political reputation of President George W. Bush, and had cost his party control of Congress in the 2006 midterm elections. In the summer of 2008, Hurricane Gustav forced the delay of the Republican National Convention, not because it was in the path of the storm but because the party wanted to avoid any reminders of the debacle that had followed Hurricane Katrina 3 years before. As *Guardian* reporter Ewen MacAskill observed, this was "the first time in living memory that a Republican or Democratic convention has been disrupted by a natural disaster" (MacAskill 2008). In January of 2009, Barack Obama was inaugurated after handily winning the popular vote in the 2008 election. In addition to being the first African-American president, he was also the first president to have made addressing climate change a major tenet of his campaign. With President Obama's support, the House passed the American Clean Energy and Security Act in the summer of 2009. This act would have established a mechanism for emissions trading in the United States similar to the one that had been established in the European Union. Unfortunately, the bill died in the Senate as opponents used procedural measures to prevent it from ever coming to a vote (Lizza 2010).

When I reached the ripe age of 50 in the summer of 2014, the earth's carbon concentration had reached a stunning 401 parts per million. The chart [Fig. 1] from the Scripps Institution of Oceanography shows the tremendous rise in atmospheric

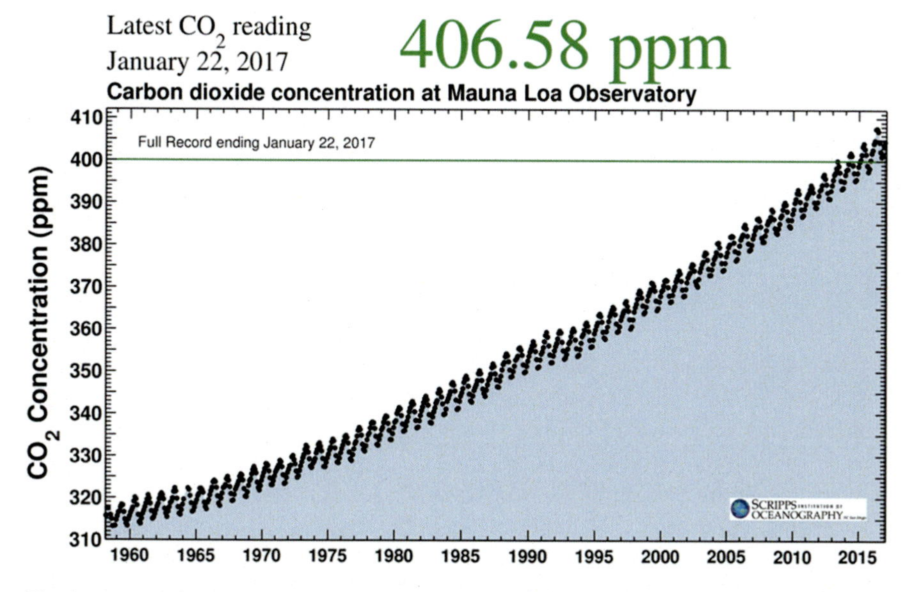


**Fig. 1** The "Keeling Curve"
*Based on measurements taken at the Mauna Loa Observatory in Hawaii, this chart illustrates the dramatic rise of CO2 in the atmosphere, measured in parts per million, since 1958*
Scripps Institution of Oceanography https://scripps.ucsd.edu/programs/keelingcurve/

carbon over the course of the past half-century. The measurements plotted on this chart were begun by Charles David Keeling at the Mauna Loa Observatory in Hawaii in 1958, and the dramatic rise in atmospheric carbon that this chart illustrates has come to be known as the "Keeling Curve." It is a powerful illustration of how radical this change has been. Fifty years ago, when talk of "space age" technology was ubiquitous, television shows such as *Star Trek* featured their characters having adventures on other planets. In the second decade of the twenty-first century, we have all become characters in a drama that takes place on another planet. Bill McKibben has called this altered world "Eaarth" and warns that its climate will not be as hospitable as the world that our ancestors have known for the past ten thousand years (2010).

The last time in the history of our planet that the level of carbon in the atmosphere had been over 400 parts per million was the Pliocene Epoch, when sea levels were about 10 m higher than they are today, and the Arctic was free of ice. Not surprisingly, the return of such atmospheric conditions has led to rapid changes in the Arctic and on Antarctica, and these are yielding extreme weather, flooding, and droughts in other parts of the earth's climate system. In turn, these changes are fueling mass extinctions across the world and play a role in such human events as wars, insurgencies, and mass migrations. The question of whether democracy, a system of government that was born and evolved during the relative stability of the Holocene epoch, can survive the coming disruptions of the Anthropocene remains an open one. If we lose democracy, we will lose a system of government that, because it allows the free exchange of information and ideas, is indispensable to the practice of science. We will also lose a system of government that, because it allows voters the chance to replace corrupt or feckless leaders with regular elections, is uniquely suited to meeting the challenges of a rapidly changing world. In other words, we face the choice between a vicious or a virtuous cycle. If we allow the disruptions caused by climate change to frighten us into embracing authoritarianism and abandoning democracy, we will be less prepared to prevent or mitigate the further disruptions that await us down the road. On the other hand, if we make a systematic effort to strengthen democracy across the world, we will be better prepared to understand and respond intelligently to the future that climate change has in store for us. In order to pursue the latter course, it will be necessary to create new democratic institutions that can operate on a global scale.

Several years ago, when the ecologist Robert Cabin approached me about writing a book for his series Environmental Challenges and Solutions, he noted that the books in this series would be distinguished from most academic writing in this simple but essential way: these books must not only describe environmental problems but also point to a plausible path for solving those problems. For me, this was a surprising proposition. Historians are not in the business of proposing solutions to problems. In fact, we are usually occupied with analyzing how past solutions have fallen short, failed, or produced unexpected consequences. Because I have chosen to discuss the issue of global climate change, I must acknowledge at the outset that envisioning a viable solution to that problem is especially challenging. As a historian, my approach has been to sift through evidence from our past in the hope of finding clues about what our options might be for the future. Throughout, my thinking has been guided by Voltaire's observation that "The perfect is the enemy of the good." The goal of constructing democratic institutions beyond the nation state should not be viewed as a millennial crusade to establish a global utopia. Rather, it should be seen as a recognition that democratic federalism has served human needs fairly well in the past and, if it is given a chance to work within the wider arena of world affairs, is likely to be a useful tool in coping with the challenge of climate change.

Because this book is concerned with how we might find a lasting solution to the challenge of climate change, the analysis presented here cannot be what Max Weber called *wertfrei*, or free of value judgments. It may be that science, as something that human beings do, can never entirely escape the influence of value judgments. In fact, the pursuit of the truth itself through the cooperative process of science is predicated on two implicit value judgments: first, that the truth itself is worth knowing, and, second, that it is worth sharing. Science, instead of being value free, is a moral quest to *face* the truth, however much that truth might upset or offend our most cherished prejudices, and it is a social quest to *share* the truth, however upsetting it might be to the existing social, political, and economic order. So, although scientists must strain to be *wertfrei* as they assess the data before them, they are engaged in a moral endeavor of truth telling that frequently has profound consequences for the human race. The social and political earthquakes that resulted from the scientific discoveries of Galileo, Darwin, and Einstein illustrate this fact. The discovery of anthropogenic climate change, which has been the work of countless scientists over the course of the past century, is producing a similar tectonic shift that has rattled the social and political order of the industrialized world.

The physicist and historian of science Timothy Ferris has documented the social and political upheavals created by new scientific discoveries in his magisterial work *The Science of Liberty*. Ferris makes the argument that science, far from being politically neutral, has a natural affinity with democracy. Because the scientific pursuit of truth demands intellectual freedom, the sciences will always thrive best in free societies. Because science necessarily bases its conclusions on evidence rather than authority, the practice of science must inevitably oppose dogmatic thinking and authoritarian practices of all kinds. In the last chapter of *The Science of Liberty*, Ferris concludes that the animating values of science and democracy are not only indispensable to each other, but will also be indispensable to addressing the challenge of climate change. When it comes to the nation state, Ferris takes the world as it is and argues that individual national governments should embrace both science and democracy in order to advance their GDP and their position in the world. The history of nation states in the twentieth century provides ample evidence to support a causal link between a vibrant democracy and the sort of scientific and technological innovation that produces wealth and power. On the other hand, science and democracy are both much older than the nation state and there is reason to hope that, if our species survives, they will continue to be human values long after the trappings of militaristic nationalism have been relegated to museums.

To date, much of the discourse on climate change has been more critical of capitalism than nationalism or authoritarianism. Naomi Klein has argued that we cannot address the problem of climate change "without challenging the fundamental logic of deregulated capitalism" (2015, p. 24). James Hansen has argued that applying market-based solutions to climate change, such as the sale of "carbon offsets," is not only ineffective but corrupt, comparing such financial devices to "indulgences that were sold by the church in the Middle Ages" (Hansen 2009, p. 206). There are many reasons to take a dim view of how self-serving economic interests have thwarted our ability to deal with climate change. However, any attempt to change the economic system of a given country must contend with the Westphalian system of state sovereignty in which that country struggles to survive. For example, the history of the most stridently socialist governments in the twentieth century indicates that, as they attempted to survive and function within the framework of sovereign nation states, they had an environmental record that was frequently far worse than that of other industrial countries. The Union of Soviet Socialist Republics, founded in 1917, and the People's Republic of China, founded in 1949, were each seriously committed to creating economic growth without capitalism, but both regimes produced some of the largest environmental catastrophes of the twentieth century. The totalitarian practices of both regimes contributed to these catastrophes, and, in a phenomenon that transcends the divisions of left and right, these practices were justified by the demands of national security.

The most environmentally destructive regimes have tended to be the most authoritarian, and authoritarian practices, whether on the left or on the right, have usually been introduced in the name of protecting that sacred abstraction known as the "national interest." This suggests that the more fundamental problem is not capitalism, but nationalism, and the authoritarianism that it engenders. The question at the heart of this study is a simple one: How can we liberate the cooperative power of democracy from the anti-cooperative institution of the nation state? How can we disentangle one of the best ideas the human race has ever had from one of its worst? In the twentieth century, this was a question that many people asked as they witnessed the tendency of nationalism to erode democracy and lead the world into catastrophic warfare. In the twenty-first century, we must ask this question again as we see nationalism eroding democracy, thwarting global cooperation to protect the environment, and dragging the world into a new epoch of catastrophic climate change.

When Al Gore received the Nobel Peace Prize for his work on climate change, he cited an old African proverb that says, "If you want to go quickly, go alone. If you want to go far, go together" (Gore 2007). Authoritarian regimes, in which the government operates alone and without the consent of the people, have often achieved short-term goals with stunning rapidity. Some dramatic examples of this have been the Qin Emperor's completion of the Great Wall of China over two thousand years ago, or Stalin's brutal industrialization of the Soviet Union in the 1930s. Such regimes can produce a temporary obedience that is based on terror, and thus command the labor of the public to achieve immediate goals, usually of a militaristic nature. However, authoritarian regimes can produce neither genuine allegiance nor lasting cooperation. These can only be achieved and maintained through democratic institutions. The challenge of climate change demands our lasting cooperation on a global scale, and so it requires the creation of a supranational democracy. The first step in that direction will be to build an association that unites the efforts of the democratic governments on every continent to protect the extraordinary gains that democratic governance has made around the world over the course of the past two centuries.

## Bibliography for Preface

Ash, Timothy Garton. 2008. We friends of liberal international order face a new global disorder. *The Guardian*. September 10th, 2008.

Bacevich, Andrew. 2008. *The limits of power: The end of American exceptionalism*. New York: Henry Holt & Company.

Bentele, K., and O'Brien, E. 2013. Jim crow 2.0? Why states consider and adopt restrictive voter access policies. *Perspectives on Politics*, 11(4): 1088–1116. https://doi.org/10.1017/S1537592713002843.

Bowman, Durrell. 2016. *Experiencing Peter Gabriel: A Listener's companion*. New York: Rowman & Littlefield.

Duchene, Francois. 1994. *Jean Monnet: The first statesman of interdependence*. New York: W.W. Norton & Co.

Emerson, Sarah. 2017. My world's on fire: We asked Smash Mouth if 'All Star' is about climate change. *Motherboard*. https://motherboard.vice.com/en_us/article/qkqdm7/is-smash-mouth-all-star-about-climate-change-global-warming. Accessed 24 Jan 2018.

Foa, R.S. and Mounk, Y. 2017. The signs of deconsolidation. *Journal of Democracy* 28(1): 5–15. Johns Hopkins University Press.

Ferris, T. 2011. *The science of liberty: Democracy, reason, and the Laws of nature*. Harper Perennial.

Gore, Al. 2007. Nobel prize lecture https://www.nobelprize.org/nobel_prizes/peace/laureates/2007/gore-lecture_en.html. Accessed 3 Jan 2018.

MacAskill, Ewen. 2008. Hurricane Gustav: Republican convention thrown into Chaos. *The Guardian*, 31 August, 2008. https://www.theguardian.com/world/2008/sep/01/usa.republicans2008

Hansen, James. 2009. *Storms of my grandchildren: The truth about the coming climate catastrophe and our last chance to save humanity*. New York: Bloomsbury.

Huntley, James R. 2001. *Pax Democratica: A strategy for the 21st century*. Palgrave.

Ikenberry, H. John and Anne-Marie Slaughter. 2006. *Forging a world order of liberty under law: Final paper of the Princeton project on National Security*. Princeton: The Woodrow Wilson School of Public and International Affairs at Princeton University.

Jeffers, Robinson, Tim Hunt, ed. 1991. *The collected poetry of Robinson Jeffers: Volume five textual evidence and commentary 1st Edition*. Palo Alto: Stanford University Press.

Klein, Naomi. 2015. *This changes everything: Capitalism vs. the climate*. New York: Simon & Schuster.

Kristof, Nicholas. 1993. China sees 'Market-Leninism' as way to future. *The New York Times*. September 6th, 1993.

Lizza, Ryan. 2009. As the world burns: How the Senate and the white house missed their chance to deal with climate change. *The New Yorker*. October 11th, 2010.

Lynch, M. 2001. Pandora's ballot box: Comments on the 2000 US presidential election. *Social Studies of Science* 31(3): 417–419. Retrieved from http://www.jstor.org.ezproxy.bu.edu/stable/3183007

McKibben, Bill. 1989. *The end of nature*. New York: Random House.
Monnet, Jean. 1978. *Memoirs*. Trans. Richard Mayne. New York: Doubleday & Company.
Mounk, Yoscha. 2018. *The people vs. democracy: Why our freedom is in danger and how to save it*. Cambridge, MA: Harvard University Press.
Mueller, John. 2001. *Capitalism, democracy, and Ralph's pretty good grocery*. Princeton: Princeton University Press.
Oreskes, Naomi & Erik M. Conway. 2011. *Merchants of doubt: How a handful of scientists obscured the truth on issues from tobacco smoke to global warming*. New York: Bloomsbury.
Pinker, Steven. 2018. *Enlightenment now: The case for reason, science, humanism, and progress*. New York: Viking.
Peterson, T.C., W.M. Connolley, and J. Fleck. 2008. The myth of the 1970s global cooling scientific consensus. *Amer Meteor Soc* 89: 1325–1338, https://doi.org/10.1175/2008BAMS2370.1
Pomeranz, Kenneth and Steven Topik. 2006. *The world that trade created: Society, culture, and the world economy*. London: M. E. Sharpe.
Poole, Robert. 2008. *Earthrise: How man first saw the earth*. New Haven: Yale University Press.
Ricks, Thomas. 2017. *Churchill and Orwell: The fight for freedom*. New York: Penguin Press.
Rosenboim, Or. 2017. *The emergence of globalism: Visions of world order in Britain and the United States, 1939–1950*. Princeton University Press.
Schulman, Bruce. 2001. *The seventies: The great shift in American culture, society, and politics*. New York: Simon and Schuster.
Scripps Institution of Oceanography. 2017. Keeling curve. January, 2017. https://scripps.ucsd.edu/programs/keelingcurve/. Accessed 23 Jan 2018.
Weart, Spencer R. 2008. *The discovery of global warming*. Cambridge, MA: Harvard University Press.

# Acknowledgments

Writing is a lonely occupation, but it is not something you can do alone. Any book requires the work of countless individuals before it can find its way to press, and this book is no exception. I cannot adequately thank every person who helped me to research, write, and revise this book, so this partial list will have to do: Robert Cabin, who approached me in 2012 about doing a book for his series Environmental Challenges and Solutions and who has offered me his warm support and honest counsel every step of the way; Donald Worster, Spencer Weart, and Robert Wexelblatt for reading earlier drafts of this manuscript and offering their kind support and frank evaluations; Mary Cleave, Russell Schweickart, Andreas Bummel, and Tiziana Stella, who took time out of their busy schedules to be interviewed for this project; the American Historical Association, the World History Association, and the Aldous Huxley Society, at whose conferences I was able present and obtain valuable feedback much of the research presented here; Judith Terpos, who has been so patient with me throughout the entire process of producing this book; Betsy Baker, who shared with me many valuable insights about the life and work of Elisabeth Mann Borgese; Chase Gorland, Isabel Keogh, and Morgan Ashurian, who were superb research and editorial assistants; Daniel Flesch, whose aid with organizing my research and citations was nothing less than indispensable; and the Center for Interdisciplinary Teaching and Learning at Boston University for supporting the work of my research assistants. Finally, I would like to thank my wife, Isadora Deese, for her love, support, and wise advice.

# Contents

# Chapter 1
# The Frontiers of Democracy

*"But I tell you the New Frontier is here, whether we seek it or not."*

John F. Kennedy

**Abstract** As the effects of climate change become increasingly disruptive in this century, the democratic nations of the world will face heightened stress from extreme weather, flooding, droughts, and mass migrations. If democratic governments respond to these challenges by embracing nationalism, they will weaken the universal principle of human rights upon which democracy is founded, thus eroding the strength of democracy within their own borders. Conversely, if democratic societies foster greater political integration with other democracies around the world, they will be in a far better position to face the transnational challenges posed by climate change.

**Keywords** Atomic bomb · World federalism · Albert Einstein · Emery Reves

In his 1960 presidential campaign against Richard Nixon, John F. Kennedy introduced the slogan "the New Frontier" to define his campaign and administration. Because President Kennedy would soon call upon the United States to send a man to the moon before the end of the decade, the concept of the New Frontier has most commonly been associated with space exploration. A few years after the shock Kennedy's assassination in the autumn of 1963, the popular television program Star Trek offered its own variation of the lost president's bold vision with the dramatic slogan, "Space: the final frontier." The ultimate success of Kennedy's Apollo program in the summer of 1969, has solidified the association between his concept of the New Frontier and space exploration ever since, in spite of the fact that the legislative agenda that Kennedy advanced under the aegis of the New Frontier was the most wide-ranging set of reforms since the New Deal, and entailed everything from conservation and aid to farmers to the reduction of tariffs and the introduction of the



R. S. Deese, *Climate Change and the Future of Democracy*, Environmental Challenges and Solutions 5, https://doi.org/10.1007/978-3-319-98307-3_1

Civil Rights Act (Bernstein 1993). Although Kennedy would not live to see the first color photographs of the earth from the surface of the moon, the power of those images, affirmed many of the themes that he sounded as president, such as the fragility of life on the world we share, and the interdependence of all nations. On July 4th, 1962 Kennedy gave a speech at Independence Hall, where both the Declaration of Independence and the Constitution had been drafted. Kennedy argued that the latter document had a special relevance for an increasingly integrated world, because, "it stressed not independence but interdependence — not the individual liberty of one but the indivisible liberty of all" (Kennedy 1962).

The idea that an interdependent world required the universal establishment of constitutional democracy was over a century old in Kennedy's time, but it had acquired a new urgency with the creation of the first atomic bombs. In the mid-twentieth century, the catastrophic threat of nuclear war inspired a number of prominent scientists and intellectuals to advance the idea that democracy can and must be practiced on a global scale, through the creation of new and accountable institutions transcending the parochial boundaries of the nation state. In 1946, the Federation of American Scientists published *One World or None* to advance the essential argument that nuclear weapons were a "world problem" for which there could "be no merely national solutions." (Masters and Way, Lippmann 1946, p. 216). The most prominent scientist in their ranks, Albert Einstein, argued that the only reliable way to control the danger posed by this new generation of weapons would be the establishment of world law through the creation of a democratically accountable "world government" (Rowe and Schulmann, eds. Rowe and Schulmann 2007, p. 373). At the time, many public figures in both the United States and the Soviet Union criticized Einstein's proposal as naïve and irresponsible (Isaacson 2007, pp. 496–497). However, more than half a century later, his idea deserves a serious reassessment. In the early twenty-first century, the challenge of climate change has strengthened the case for the expansion of democracy beyond the nation state and the implementation of democratic accountability on a global scale.

As we confront the catastrophic dangers posed by climate change, there are three reasons why this unprecedented crisis demands the creation of a global democracy. The first reason is that we cannot curb global carbon emissions without a supranational carbon tax. A carbon tax is necessary to make the planetary cost of greenhouse gas production a consistent factor in the economic decisions of both producers and consumers. As James Hansen puts it, "the essential backbone" of any effective path to dealing with climate change requires "a rising price of carbon applied at the source (the mine, the well-head, or port of entry) such that it would affect all activities that use fossil fuels, directly or indirectly" (Hansen 2009, p. 205). However, unless the imposition of a cost on carbon is global, it cannot be effective. This brings us to the second reason for the creation of a global democracy: a supranational carbon tax requires a global governing structure to collect that tax, and to levy fines or other economic penalties for its evasion, and to determine how the revenues from the tax will be invested. These issues underline the third and most essential reason for a global democracy: a supranational carbon tax must not be a form of taxation without representation. In other words, a global carbon tax cannot be legitimate unless it is levied by a representative assembly, practicing democracy on a global scale. Even if its initial purpose was simply to collect and invest a carbon tax, the creation of a supranational legislature that spanned the globe would be an unprecedented step in the political history of the human race.

Since the year 2000, the conviction that global democracy must be part of the solution to climate change has found support from a growing number of scholars, journalists, and political activists. In *Democracy and Global Warming* the political scientist Barry Holden draws on Aristotle to make the case that democracy can marshal knowledge and expertise dispersed throughout a population and is therefore better at dealing with unexpected challenges (2002, p. 44). In many ways, Holden's scholarship builds on that of David Held and Daniele Archibugi, who have been exploring the possibility of extending democracy beyond the nation state since the late twentieth century (Archibugi and Held 1995). Like these scholars, Holden makes a detailed case for cosmopolitan democracy, but expands upon their thinking by arguing that the creation of a democracy without borders would offer indispensable advantages for dealing with climate change.

Some veteran scholars of international law, such as Richard Falk and Andrew Strauss, have asserted that compliance in international agreements will continue to be haphazard and prone to failure until we create institutions with sufficient democratic legitimacy to override traditional appeals to national sovereignty: "Only when the veil of sovereignty is pierced by a global parliament deriving its authority directly from citizens will there be an organization potentially capable of exerting some supranational control over states" (Falk and Strauss 2002). Pointing to the gradual development of the European Parliament, they argue that a global parliament could "begin life as a largely advisory body" and later gather strength, as the European Parliament has. They reason that, "with the power of popular sovereignty behind it, the Global Parliament would, like the European Parliament, grow in power and stature over time" (Falk and Strauss 2002). Observing that the "European Union began as a group of only six nations," Falk and Strauss argue that a "directly elected body with universal aspirations comprised of only 20 to 30 founding members would constitute an impressive start and could make a significant impression" (2002). Their arguments join those of other advocates who make the case that a global parliament could "challenge the ability of states to opt out of collective efforts to protect the environment" (Bray and Slaughter 2015, p. 83).

Among journalists, the *Guardian* columnist George Monbiot has made the most extensive and impassioned case for global democracy in his book *The Age of Consent: Manifesto for a New World Order* (2003). Arguing that economic globalization must now be balanced by the globalization of democratic institutions, Monbiot emblazoned his book with the slogan, "Everything has been globalized except our consent." To address this democratic deficit, Monbiot advocates the creation of a world parliament whose resolutions would be non-binding at first, but which would provide a democratic counterweight to the policies of the WTO, IMF, and other economic organizations. As this parliament gained in moral authority and influence, it could take on incrementally greater responsibilities, much as the House of Commons has over the course of its history in Britain. Addressing economics directly, Monbiot also advocates some of the global economic reforms that John Maynard Keynes proposed (and saw rejected) at Breton Woods, such as a global reserve currency to prevent any one nation from obtaining unfair economic advantages over its trade partners (Monbiot 2003, pp. 160–164).

Among political activists, the entrepreneur and internet activist Peter Schurman has recently emerged as a prominent supporter of global democracy. In 1998, Schurman was a co-founder of the liberal activist website MoveOn.org, which would prove to be one of the most effective political websites in the United States during the early twenty-first century. His new vehicle for internet activism, OneGlobalDemocracy.com is a forum for arguments in favor of a single, democratic world government and the case for greater environmental sustainability figures prominently among those arguments. Profiled in the business magazine *Fast Company* in 2017, Schurman argues that innovations such as the BlockChain technology pioneered by cryptocurrency developers could create a reliable platform for global democracy. Although his ideas are incredibly ambitious, Schurman emphasizes that they are grounded in what he sees as basic common sense: "Fundamentally, the idea of working together as a global community to solve the common problems we all face makes so much sense that once people start to really consider it, it's just kind of a no-brainer" (Peters 2017). Working along parallel lines, the international group Democracy Earth has been working since 2013 to design voting systems using BlockChain technology in Argentina and other countries (Peters 2016).

Of course, there are also those who make the opposite case, promoting the idea that the current ecological crisis necessitates a retreat from democracy. In their book *The Climate Change Challenge and the Failure of Democracy* (2007), David Shearman and Joseph Wayne Smith present the argument that liberal democracy has proven itself incapable of dealing with climate change. Where Barry Holden has cited Aristotle to argue for the virtues of democracy, Shearman and Smith cite Plato to expound its vices. Arguing for the creation of technocratic authoritarian governments that would in many respects resemble Lee Kuan Yew's Singapore, they also imagine the creation of a new religion based on ecological sustainability and the training of "eco warriors" to spread its gospel. Shearman and Smith fault George Monbiot for making the naïve assumption that "all will be well with more democracy and a world parliament" (2007, p. 121). Beyond oversimplifying Monbiot's *Age of Consent*, these authors make some broad and unsupported assumptions when they imagine a government of enlightened technocrats who will never need to be held accountable by such mechanisms as a free press, an independent judiciary, or competitive elections. While their critique of contemporary democratic governments is often cogent, the authoritarian solution they propose leaves a number of questions unanswered, not the least of which being the question raised by the Roman poet Juvenal: "Who will guard the guards?"

The most useful reflections on the direction of democratic societies in the twenty-first century have come from those political scholars who have been less categorical in their critiques of democracy. While they acknowledge its record of success in the twentieth century, these scholars are ready to analyze the weaknesses that have challenged democracy in the past and are likely to do so again in the future. In his book

*Democratic Faith* (2005), the conservative Catholic scholar Patrick J. Deneen argues that faith in democracy has too often been tied to a misguided belief in the perfectibility of human nature. Inspired by the theologian Reinhold Niebuhr's critique of utopian progressivism in the mid-twentieth century, Deneen contends that we should support democracy precisely *because* human beings will always be flawed, and no individual or group of individuals should ever be trusted with unchecked authority. Niebuhr had originally lent his support to the Committee to Draft a World Constitution at the University of Chicago in 1945, but he soon left the world federalist movement after concluding that its goals were too idealistic to be applicable to the challenges of the Cold War (Rosenboim 2017, pp. 174–176). However, he still shared a faith that democratic institutions were vital for humanity. As Niebuhr put it in *The Children of Darkness and the Children of Light*, "Man's capacity for justice makes democracy possible, but man's capacity for injustice makes democracy necessary" (Deneen, p. 255). Deneen ties this quote to debates about the human capacity for self-government that extend back for centuries. Supporting the faith in our "capacity for justice," he hears the voices of "Jefferson, Emerson, Whitman, and Dewey, among others." Warning about our "capacity for injustice," he hears the voices of "Hobbes, Locke, Madison, and Hamilton" (Deneen, p. 255). As Deneen sees it, both are telling us part of the truth, but if we ignore those voices that warn us about inherent human limitations, we will put democracy in peril.

Writing from a somewhat more liberal perspective, the political scholar David Runciman has also warned against overconfidence by the advocates of democracy. In his book *The Confidence Trap* (2013), Runciman argues that democracies have proven remarkably resilient in dealing with sudden and overwhelming crises, but that they have a poor record of dealing with systemic and gradual crises (Runciman, p. 317). Because advocates of democracy can point to situations in the past, such as the Great Depression and World War Two, in which democracies have survived dramatic catastrophes and ultimately prevailed, they tend to be complacent in the face of less conspicuous crises, especially if they stem from problems that become worse incrementally. It is this complacency in the face of incremental problems that Runciman labels "the confidence trap," and he sees this danger as being especially relevant to the environmental challenges of this century. Although he acknowledges that the environmental record of democracies is superior to that of autocratic regimes, he worries that the confidence born of success in previous crises will lead democratic governments to procrastinate in the face of climate change until effective action becomes impossible (Runciman, pages 315–317). Surveying recent environmental debates in the United States, Runciman concludes, "That is why climate change is so dangerous for democracies. It represents the potentially fatal version of the confidence trap" (Runciman, 318). However, because Runciman surveys the history of national governments, he can only evaluate the record of democratic governments that have been elected to serve the "national interest" of their respective electorates. The only supranational parliament in the world is that of the European Union, and its record on climate change has been among the most effective of any governing body in the world (Rayner and Jordan 2016). In relation to climate change, there is reason to believe that the shortsightedness inherent in nationalism has been at least as great an impediment to effective action as the complacency that Runciman sees as inherent in democracy. When democratic societies such as the United States have failed to act responsibly on climate change, as in George W. Bush's rejection of the Kyoto Accords in 2001 or Donald J. Trump's rejection of the Paris Agreement in 2017, their failures have had less to do with democracy than with nationalism. While neither of these leaders had won the popular vote in the elections that first brought them to power, both of them employed the stark rhetoric of nationalism when they reversed the climate policies of their predecessors.

Whatever one's political persuasion, it is wise to be aware of the fallibility of any political system, and democracy is no exception. One of the primary virtues of democracy, however, is its ability to respond to its own fallibility. In this sense, the process of democratic governance resembles the process of scientific inquiry. The practice of peer review that defines science is predicated on the idea that we should never accept any premise merely on authority, but always see for ourselves. In a similar manner, electoral democracies check political power with their own methods of "peer review" such as a free press, competing political parties, and regular elections. This rough parallel between science and democracy was noted by the philosopher Horace M. Kallen during the Second World War, as he organized a seminal series of conferences in New York City on "The Scientific Spirit and Democratic Faith." Kallen's conferences attracted such luminaries as the philosopher John Dewey and the political thinker Sidney Hook, who shared his conviction that the values of science and democracy had a fundamental and enduring affinity. More recently, the physicist Timothy Ferris has argued that science and democracy are not only similar but are also linked in a symbiotic relationship. In his analysis of the European history during the 17th and 18th centuries, Ferris maintains that the scientific revolution paved the way for the expansion of democracy, and that the widening influence of democratic values engendered further advances in science. As he traces their parallel histories across four centuries, Ferris observes that, like democracy, "science is inherently antiauthoritarian" requiring both freedom and transparency in order to flourish (Ferris 2011, pp. 4–6). Ferris also points to the data compiled by odds-makers and futures markets around the world supporting the hypothesis that groups of individuals "can demonstrate superior predictive power" especially when the individuals comprising those groups have a stake in the outcome and are allowed to function, not as a mob, but "independently of one another, making their own determinations rather than being influenced by others" (Ferris 2011, p. 32). Ferris notes that "the accuracy of group predictions improves with its diversity: The more socially, ethnically, sexually, and intellectually diverse the group, the better it performs" (Ferris 2011, p. 32). The genius of democracy as practiced on a local and national scale has helped human beings to build societies that are "more prosperous and peaceful" than their predecessors (Ferris 2011, p. 33). Now that science and technology have endowed humanity with the ability to transform its environment on a global scale, we are consequently faced with the question of whether we can harness the power of democracy to provide its proven benefits of creativity, accountability, and resilience on a global scale.

The future of our species in the age of climate change will depend on how we answer two basic questions. First, are we capable of governing ourselves? Second, are we capable of using the fire that technology has put into our hands without burning up the world? The first question concerns our relationship to each other, while the second question concerns our relationship to nature. If there was ever a time when the two questions could be answered separately, that time has long since passed. As it stands now, these questions must be faced together, and the answer will be the same for both. If the answer to these questions is no, it will be a definitive no, settled once and for all time. This negative result will most likely entail a collapse of the world's major democracies and the ecologically stable systems on which they depend. If the answer to these questions is yes, it will always be a provisional yes, as when we say "so far, so good" and keep a wary eye on the future.

We clearly need all the help we can get, so it makes sense to draw whatever wisdom we can from the past. Like Pandora's Box, the cultural and historical legacy of Ancient Greece is a gift that keeps on giving. Greek mythology and history provide the most compelling tales about the origin of both fire and democracy. The mythical figure of Prometheus has proven to be a durable symbol for the tremendous promises and risks associated with modern science, while the historical figure of Cleisthenes is still revered as one of the founders of Athenian democracy. In an age of climate change and political globalization, Prometheus and Cleisthenes still have much to teach us. Hesiod's *Theogeny* tells the story of how humans, inspired by the titan Prometheus, tricked the gods into accepting bones covered with animal fat as a counterfeit for the customary "burnt offerings" of the best meat at the table. When the gods realized that they had been swindled by appearances, they denied the use of fire to humans. To set things right again, Prometheus stole fire from the gods and gave it back to the human race. For his crime against the power and privilege of the gods, Prometheus was summarily chained to a rock so that an eagle could feast upon his liver. When this organ, which the Greeks believed to be the seat of the emotions, had been consumed, it would grow back so that this profound agony could begin again and last for all eternity (Hesiod 2017, pp. 70–72). The name Prometheus means forethought, which suggests that this titan knew that he would face a harrowing punishment for his crimes, but, foreseeing the tremendous potential of the human race, he believed that it was worth the sacrifice. The story also foreshadows the rise of free inquiry and a diminished role for religious regimentation in human affairs. Fire had been used for strictly religious purposes, until the humans followed the lead of Prometheus and outsmarted the gods.

As Prometheus had resisted the tyranny of the gods, Cleisthenes resisted the tyranny of Isagoras, an authoritarian traitor who imported troops from Sparta to tighten his control of Athens. The popular revolt that brought Isagoras down required a realignment of Athenian politics, and Cleisthenes made sure that this new political order would include the enfranchisement of all who ranked as Athenian citizens. Because of its dependence on maritime trade and naval power, Athens had already extended "many rights of participatory democracy" to male citizens of modest means, as their muscle power was essential to the success of the Athenian fleet of trireme warships (Kaplan 2015, p. 8). Nonetheless, a complex web of local power structures still obstructed the growth of democracy in the city state. As he surveyed the powerful local elites that controlled the various regions within Athens, Cleisthenes recognized that local and tribal identities are antithetical to the expansion of democracy. That is why he created a system of *demes* that transcended the regional and tribal affiliations that had dominated Athenian political culture. Because the *demes* drew on people from each of the regions and tribes of Athens, they lessened the influence of group identity and encouraged Athenians to think of themselves as citizens of the entire polis first. While nothing can eliminate the proclivity of human beings to see themselves as partisans of a particular region, tribe, or walk of life, the reforms of Cleisthenes at least reduced that proclivity in order to advance Athenian democracy (Buckley 2010, pp. 125–127).

The combined legacy of Prometheus and Cleisthenes has come to shape our world. Once humans began to control fire, they began the process of transforming the world around them. Once the first human societies experimented with the creation of democratic institutions, they began to learn, in fits and starts, how larger communities of individuals could break free from the bonds of arbitrary authority and learn to govern themselves according to the principles of reason. In an age when multinational corporations and other non-state actors have long since transcended national borders, the concept of "national sovereignty" has already become more myth than reality. At the beginning of the twenty-first century, the political scientist David Held observed that multinational corporations already accounted for about seventy percent of world trade and eighty percent of foreign direct investment, with "profound effects on macroeconomic policy" by shifting "demand for employment to countries with much lower employment costs" and moving all economic and technological enterprises to wherever "maximum benefits accrue"; with the natural consequence that, "the autonomy of democratically elected governments has been, and is increasingly, constrained by sources of unelected an unrepresentative economic power" (Holden, ed. 2000, pp. 20–22). Efforts to reassert the primacy of the nation state through scapegoating immigrants or pursuing a zero-sum foreign policy may offer a temporary catharsis to some, but they are destructive to democratic values at home and effective cooperation abroad. As the effects of climate change, such as dangerous heatwaves, wildfires, droughts, and violent weather events proliferate in the coming century, stresses on both democratic values and cooperation across national boundaries are likely to become more severe. Some democratic societies may sever the bonds of cooperation with their neighbors and even embrace the dubious promise of authoritarian rule. If democracy is to survive, particularly on a planet shaken by the violent disruptions of climate change, it must grow beyond the borders of the nation state, and ultimately embrace the entire planet. Only with the creation of global institutions that are disciplined by democratic accountability will it be possible to check the power of multinational corporations, address the causes of climate change, and mitigate its effects.

In our own time, the group identity of nationalism, though it has some virtues, has become another impediment to the evolution of democracy. Just as Cleisthenes found a way to transcend local identities to create a sense of common purpose for

Athens, we must do the same on a global scale if we are to meet the challenge of climate change. Applying the democratic vision of Cleisthenes on a global scale would be a quantum leap in the evolution of our political institutions, but by the mid-twentieth century it was an idea even the most skeptical philosophers had begun to take seriously. In *The Open Society and Its Enemies*, Karl Popper addressed the possibility of a democratic world government in some detail, reasoning that "many things have been realized which have once been dogmatically declared as unrealizable, for instance, the establishment of institutions for securing civil peace, *i.e.*, for the prevention of crime within a state." Building on this precedent, Popper argued that, "the establishment of corresponding institutions for the prevention of international crime, *i.e.*, armed aggression or blackmail, though often branded as Utopian, is not even a very difficult problem" (Popper 1962, p. 161). Popper not only argued that a supranational government with police authority for the prevention of war and other forms of aggression was an eminently practical goal, but he also insisted this sort of institution could be crafted in such a way that it provided a surer guarantee of individual rights than the existing system of sovereign nation states. In cases where the proposed supranational authority had to redress an act of aggression by a particular nation state, Popper adamantly insisted that, "We must realize that *we can treat individuals fairly even if we decide to break up the power-organization of an aggressive state* or 'nation' to which the individuals belong" (Popper 1962, p. 289. Italics in original). Popper's assertion that the creation of a constitutional order beyond the nation state "is not even a very difficult problem" must of course be taken with a grain of salt. At the very least, however, it should be acknowledged that the concept of a democratic world government meets Popper's basic standards for a *falsifiable* proposition, and is therefore worthy of serious consideration. Unlike such dubious notions as "national greatness," which can never be clearly defined or tested in the real world, the prospect of holding valid elections and enforcing laws on a transnational scale has already been tested by the European Union, and could plausibly be tested on a much larger scale.

Popper had characteristically modest expectations about the ability of any sort of world government to prevent war entirely, and he reasoned that the League of Nations had been woefully ineffective at responding to military aggression in the 1930s because of the unrealistic expectation that "the League had been established in order to end all wars and not to wage them" (Popper 1962, p. 288). Although he did not expect that a supranational government would end all violent conflict in the world, he did expect that it could protect the rights of individuals better than the shifting configuration of power centers that would come to be called "the anarchical society" of sovereign nation states (Bull 1977). Arguing that the borders of nation states were the result of historical accident, Popper reasoned that, "*human individuals and not states or nations must be the ultimate concern even of international organizations*." (Popper 1962, p. 288. Italics in original). This was a principle that was also affirmed after World War Two by figures as diverse as the physicist Albert Einstein, the political journalist Walter Lippmann, and U.S. Supreme Court Justice Owen J. Roberts (Masters and Way, Lippmann 1946; Streit et al. 1950).

By the standards Popper outlined in *The Open Society and Its Enemies*, the international structure of the United Nations was barely an improvement over the League of Nations. In contrast to the League, the UN has been able to mount a military response to some forms of international aggression, as in the case of the North Korean invasion of South Korea in June of 1950, or the Iraqi invasion of Kuwait in August of 1990. On the other hand, it has been powerless to prevent disastrous military misadventures by the permanent members of its Security Council, such as the American war in Vietnam (1965 to 1973), the Soviet invasion of Afghanistan (1979 to 1988), or the American invasion of Iraq (2003 to 2011). It has also failed to prevent ethnic genocides and other mass killings in such places as Bosnia, Rwanda, and Syria. The reasons for these failures are many and complex, but a consistent adherence to the doctrine of national sovereignty ranks high among them. The five permanent members of the UN Security Council each see their possession of the veto as essential to their national sovereignty, even though it has crippled the ability of the UN to prevent or even mitigate significant military and humanitarian catastrophes such as the brutal counterinsurgency conducted by the Russian client state of Bashar al Assad. (Cadier and Light 2015, p. 2). Under the aegis of national sovereignty, a multitude of national governments that do not possess the veto are still able to carry out policies that are catastrophic for human rights as long as they are carrying out those policies within their own borders, or remain a client state to a permanent member of the UN Security Council.

The record of the UN since 1945 supports Popper's argument that an effective supranational organization should exist to protect the rights of individual human beings, not national governments. Whereas the preamble to the United Nations begins with the phrase, "We, the Peoples of the United Nations," the preamble to a fully democratic supranational organization would begin with the words, "We, the people of the planet Earth." This latter construction is many times more ambitious than the former, but it has freed itself from the burden of two perilous abstractions, i.e., "peoples" and "nations." If the "peoples" of the world are distinct entities, what marks the distinction between them? Is it ethnicity, religion, language, or something else? And how does the distinction between "peoples" relate to the distinction between "nations," especially when some nation states are relatively homogenous in terms of ethnicity while others are incredibly diverse? "People," as the straightforward plural of "person," is the only stable concept on which any lasting democracy can be built. Mercurial abstractions such as "peoples" or "nations" cannot provide a stable foundation for any democracy. In 1918, Woodrow Wilson grandly pronounced in his Fourteen Points speech that, "All the peoples of the world are in effect partners" in seeking a new international order based on "justice and fair dealing" (Wilson 1918). His choice of the abstract concept of "peoples" over the more concrete and granular term "people" wedded his thinking to the abstract rights of nations and ethnic groups rather than the concrete rights individuals. The Hungarian journalist and polymath Emery Reves, whose thinking would greatly influence the world federalist vision of Albert Einstein, was an articulate critic of this abstraction. As a Hungarian Jew who was fluent in ten languages, Reves was opposed to the virulent brand of nationalism that was on the rise across Europe. A very public critic of fascism, Reves had to flee Nazis twice himself, first after their rise to power in Berlin in 1933, and again after their invasion of France in 1940. Before he met Einstein, Reves had found his métier as a journalist committed to translating and publishing editorials by Europe's leading liberal politicians in as many countries as possible. After resurgent hysteria over national and ethnic identity dragged Europe into a second catastrophic war, Reves argued that these concepts were inherently antidemocratic. In his book *A Democratic Manifesto*, he declared, "It must be understood that there exist only two realities – the *individual* and *humanity*. All other classifications into castes, tribes, classes, religions, races and nations are arbitrary, artificial and superficial" (Reves 1942, p. 51. Italics in original.) This is a truth that speaks to the current political crisis of Western democracies, as struggles about race, religion, and identity have emerged with a vengeance. Citing the divisive role of "identity politics" in the 2016 election, the historian Mark Lilla has called for a "post-identity liberalism" that "would speak to the nation as a nation of citizens who are in this together and must help one another" (Lilla 2016). The merits of Lilla's argument were passionately debated among liberals in the United States after Trump's Electoral College victory in November of 2016, but few if any remarked upon the fact that nationalism itself is a form of "identity politics," with its own long history of myopia and irrational violence. For the near term, it remains necessary to frame political goals in national terms, as we still live under national governments. For the longer term, however, it is no less necessary to think about how we can protect such essential values as freedom of thought and freedom of religion within a framework that is more enduring, and more rational, than allegiance to the symbols, rituals, and authorities of a particular nation state.

As Emery Reves saw it, nationalism would inevitably morph into authoritarian rule when faced with the pressures of economic globalization. Reves argued that the irreducible tension between nationalism, which is inherently parochial, and modern industrialism, which looks for resources, labor, and markets all over the world, would ultimately lead to "the destruction of political democracy and individual liberty" unless people could create a new kind of democracy that transcended the nation state (Reves 1945. pp. 100–101). When Reves wrote *The Anatomy of Peace*, his argument was bolstered by the rise of fascist regimes in response to disruption of the global economy during the Great Depression. In our own time, a very similar drift toward authoritarian rule has emerged as a response to the economic disruptions that have reverberated around the world following the financial crisis of 2008. On the eve of that crisis, the historian Paul Kennedy predicted that the United Nations would face further challenges, but that it should stick "strengthening that three-legged stool of peace, development, and democracy envisaged sixty years before" by its founders (2006, p. 278). For democracy to remain effective in the promotion of peace and development, however, it will have to build new institutional bridges beyond the constricting boundaries of national identity.

However, it is necessary to concede that transcending the parochial divisions of nationalism would not guarantee freedom from all the dangers of tyranny. There still remains the question of what sort of institutions provide the best safeguard for any democratic polity, including a transnational one, against the rise of autocracy. In his discussion of government, Popper articulates the core principle of classical liberal thought when he concludes his chapter on "Leadership" with this quotation: "As Lord Acton says – all power corrupts, and absolute power corrupts absolutely." (Popper 1962, p. 137). Consequently, power must be divided if it is not to corrupt the bearer of it and devolve into an autocracy. This was a truth that Locke intuited in his *Second Treatise on Government* (1689) and which Montesquieu developed further in his *Spirit of the Laws* (1748). In both cases, it is essential to remember that the divisions they described were *within* a given government, and not among separate sovereign governments. For a citizen of a constitutional democracy, the rivalry between the distinct branches of one's government is essential to the prevention of autocracy. Conversely, the rivalry between one's own sovereign government and another sovereign government, especially if that rivalry involves the possibility of armed conflict, *increases* the likelihood of autocracy. It is wise to worry that a global government could somehow devolve into a global autocracy. However, the most effective way to address this danger is not to cling to a patchwork of sovereign nation states, which could all too easily devolve, through the process of rivalry and war, into a patchwork of sovereign autocracies. Safeguarding liberty on a global scale will require a system of supranational governance that divides power constitutionally rather than militarily. Of course, the extension of democratic accountability beyond national borders will require us to transform our practical approach to both sovereignty and democracy.

We stand before a new frontier in the history of democracy, and the ambiguity inherent in the term *frontier* expresses both the danger and the promise of this historical moment. The ancient meaning of frontier is simple to grasp: it is a border between two territories, a place for barriers, watchtowers, and fortifications. The modern meaning of the term frontier owes a great deal to a seminal address by the historian Frederick Jackson Turner at the Chicago World's Fair in 1893. Turner argued that the open, hazardous, and improvisational culture of frontier life had been essential to the evolution of democracy in America. Inverting the cliché that European settlers had mastered the western wilderness, Turner asserted that, in the realm of political culture, "The wilderness masters the colonist" (Turner 1893). By this, Turner meant that the common challenge of survival on the western frontier stripped away old distinctions of class, religion, and ethnicity and demanded that commitment to cooperation that is the foundation of citizenship in a democracy.

More than three generations later, John F. Kennedy alluded to Turner when he incorporated the concept of "the New Frontier" in his acceptance speech at the Democratic National Convention in the summer of 1960. Echoing Turner's theme, Kennedy declared that the crucible of the frontier experience had produced an indispensable alloy of courage and cooperation. The settlers who joined together in the western wilderness, "were not the captives of their own doubts, the prisoners of their own price tags. Their motto was not 'every man for himself—'–but 'all for the common cause'" (Kennedy 1960). Without a doubt, Kennedy's New Frontier speech was shaped by the tensions of the Cold War, but it described a common future that could only be achieved by tearing down old barriers, not erecting new ones. This was an idea that he brought to fruition in his "Declaration of Interdependence" address of July 4, 1962, in which he declared that the spirit of cooperation that had spanned the Atlantic since the 1940s must be maintained and extended across the world. He expressed the hope that this spirit of interdependence "would serve as a nucleus for the eventual union of all free men—those who are now free and those who are vowing that some day they will be free" (Kennedy 1962).

When Kennedy delivered his "Declaration of Interdependence" speech in the summer of 1962, the greatest fear that concerned the world was the specter of nuclear annihilation, as the Cuban Missile Crisis would dramatically demonstrate in October of that year. More than half a century later, the threat of nuclear conflict has not disappeared, and it has now been joined by the accelerating threat of climate change, which portends global instability. The solution to which Kennedy alluded, an interdependent "union of all free men," is both more plausible and more urgent in the twenty-first century than it was in the twentieth. It is more plausible because advances in communication have shrunk the world considerably since 1962, and roughly six out of ten countries across the world have become viable democracies (De Silver 2017). It is more urgent because the challenge of climate change will require sustained cooperation on a global scale, disciplined by the accountability and transparency that only democratic institutions can provide. John F. Kennedy's declaration that "the new frontier is here whether we seek it or not" is as true now as it was in his lifetime (Kennedy 1960). Whether we greet it with hope or fear, the challenges and opportunities of the future loom over us, and no wall that we might dream of building can shut them out.

## Bibliography


Archibugi, Daniele, and David Held. 1995. *Cosmopolitan democracy: An agenda for a new world order*. Boston: Polity Books.

Bernstein, Irving. 1993. *Promises kept: John F. Kennedy's new frontier*. Oxford: Oxford University Press.

Bray, Daniel, and Steven Slaughter. 2015. *Global democratic theory: A critical introduction*. Cambridge: Polity Press.

Buckley, Terry. 2010. *Aspects of Greek history 750–323 BC: A source-based approach*.

Bull, Hedley. 1977. *The anarchical society: A study of order in world politics*. New York: Columbia University Press.

Cadier, D., and M. Light, eds. 2015. *Russia's foreign policy: Ideas, domestic politics and external relations*. New York: Palgrave Macmillan.

De Silver, Drew. 2017. *Despite concerns about global democracy, nearly six-in-ten countries are now democratic*. Washington, DC: FactTank Blog, Pew Research Center. http://www.pewresearch.org/fact-tank/2017/12/06/despite-concerns-about-global-democracy-nearly-six-in-ten-countries-are-now-democratic/ (Accessed 5 Jan 2018).

Deneen, Patrick J. 2014. *Democratic faith*. Princeton: Princeton University Press.

Einstein, A., and R. Swing. 1947. *Atomic war or peace*. Emergency Committee of Atomic Scientists. In *Einstein, Out of My Later Years*. New York: The Philosophical Library, Inc. 1950.

Falk, Richard, and Andrew Strauss. 2002. "People want a say: Next, a global parliament" *New York Times*. April 19, 2002.

Ferris, T. 2011. *The science of liberty: Democracy, reason, and the Laws of nature*. New York: Harper Perennial.

Hansen, James. 2009. *Storms of my grandchildren: The truth about the coming climate catastrophe and our last chance to save humanity*. New York: Bloomsbury.

Hesiod. 2017. *The poems of Hesiod: Theogony, works and days, and the shield of Herakles* Trans. Barry B. Powell. Oakland: University of California Press.

Holden, Barry, ed. 2000. *Global democracy: Key debates*. London: Routledge.

———. 2002. *Democracy and global warming*. London: Continuum.

Isaacson, Walter. 2007. *Einstein: His life and universe*. New York: Simon and Schuster.

Kennedy, John F. 1960. "Democratic National Convention, 15 July 1960" John F. Kennedy Presidential Library. https://www.jfklibrary.org/Asset-Viewer/AS08q5oYz0SFUZg9uOi4iw.aspx.

———. 1962. "Address at Independence Hall, Philadelphia, Pennsylvania, July 4, 1962" John F. Kennedy Presidential Library. https://www.jfklibrary.org/Research/Research-Aids/JFK-Speeches/Philadelphia-PA_19620704.aspx

Kennedy, Paul. 2006. *Thwe parliament of man: The past, present, and future of the United Nations*. New York: Vintage.

Lilla, Mark. 2016. "The End of Identity Liberalism" *New York Times*. Sunday, November 18th, 2016.

Lippmann, Water. 1946. "International Control of Atomic Energy". In *One world or none*. [first published in] (New York: The New Press, 2007), 191, 195, 205.

Masters, Dexter, and Katherine Way, eds. 2007. *One world or none*. New York: The New Press.1946

Monbiot, George. 2003. *The age of consent*. London: Flamingo.

Peters, Adele. 2016. *"Democracy is getting a reboot on the blockchain" fast company*. https://www.fastcompany.com/3062386/democracy-is-getting-a-reboot-on-the-blockchain.

———. 2017. *"The case for eliminating countries and instituting a global democracy," Fast company*. https://www.fastcompany.com/3067153/the-case-for-eliminating-countries-and-instituting-a-global-democracy

Popper. 1962. *The open society and its enemies*. Volume one. Princeton University Press, 161, 289, 288, 137.

Rayner, T., and Jordan, A. 2016–08–05. Climate Change Policy in the European Union. *Oxford Research Encyclopedia of Climate Science*. Retrieved 19 Jul. 2017, from http://climatescience.oxfordre.com/view/10.1093/acrefore/9780190228620.001.0001/acrefore-9780190228620-e-47

Reves, Emery. 1942. *A democratic manifesto*. New York: Random House.

Rosenboim, Or. 2017. *The emergence of globalism: Visions of world order in Britain and the United States, 1939–1950*. Princeton: Princeton University Press.

Rowe, David, and Robert Schulmann, eds. 2007. *Einstein on politics*. Princeton: Princeton University Press.

Runciman, David. 2013. *The confidence trap: A history of democracy in crisis from world war I to the present*. Princeton: Princeton University Press.

Shearman, David, and Joseph Wayne Smith. 2007. *The climate change challenge and the failure of democracy*. Westport: Praeger Publishers.

Streit, Clarence K., Owen J. Roberts, and John F. Schmidt. 1950. *The new federalist*. New York: Harper.

Turner, F..1893. *The significance of the frontier in American history*. https://sourcebooks.fordham.edu/mod/1893turner.asp Accessed 29 May 2017.

Wilson, Woodrow. 1918. President Woodrow Wilson's fourteen points. In *The Avalon project*. Connecticut: Yale University. http://avalon.law.yale.edu/20th_century/wilson14.asp.

# Chapter 2
# Nationalism and the "End of Nature"

> *"Nationalism is an infantile disease. It is the measles of mankind."*
>
> Albert Einstein

**Abstract** An analysis of the nexus between international relations and environmental sustainability since the early twentieth century indicates that strategic competition among sovereign nation states has greatly impeded efforts to understand and address environmental challenges. In particular, the rise of militaristic nationalism has caused extensive environmental destruction and has frequently corrupted the practice of science by tying it to the secretive culture of the national security state. During the twentieth century, numerous advocates of environmental sustainability struggled to limit the destructive impact of nationalism on scientific cooperation, and their work has attained a new relevance in the age of climate change.

**Keywords** Total war · Bertrand Russell · Patrick Geddes · Lewis Mumford · Neotechnic

In the summer of 2016, the British newspaper *The Guardian* reported that a panel of scientists who had gathered at the International Geological Congress in Cape Town, South Africa had reached the conclusion that "Humanity's impact on the Earth is now so profound that a new geological epoch – the Anthropocene – needs to be declared" (Carrington 2016). While the term Anthropocene has come into fairly wide use in academic circles, it is not widely used by the general public. Expressing the same concept in more colloquial terms, the environmental writer Bill McKibben observed in the late 1980s that climate change had brought us to "the end of nature" (McKibben 1989). Elaborating on the various factors that had led to the end of nature, McKibben described our increasing emission of greenhouse gases as an unprecedented experiment in the history of the planet: "While there are other parts to this story – the depletion of the ozone, acid rain, genetic engineering – the story



R. S. Deese, *Climate Change and the Future of Democracy*, Environmental Challenges and Solutions 5, https://doi.org/10.1007/978-3-319-98307-3_2

of the end of nature really begins with that greenhouse experiment and what will happen to the weather" (McKibben 1989, p. 9).

If we are to understand the significance of the Anthropocene or the "end of nature" we must first acknowledge that the collective impact of our species upon the earth and its ecosystems has been entirely haphazard. In the starkest terms: *agency is not control*. While it is quite likely that we have had such a profound impact on the planet that we have set in motion a new geological epoch, it is far less likely that we are in any position to guide how that epoch will unfold. In all probability, the best we can do is seek to limit our impact, and to mitigate the damage we have already done. As the effects of climate change have become more conspicuous, extreme weather events have become more common and their dangerous volatility offer a preview of what is to come. Of particular concern is the problem of tipping points that are likely to accelerate climate change and unleash further volatility. For example, the loss of sea ice reduces the earth's albedo, leading to further warming of the polar regions, and, in a positive feedback loop, the loss of even more ice. As "permafrost" thaws in arctic regions such as Northern Canada, Alaska, and Siberia, methane that has been trapped for thousands of years is released into the atmosphere. Since methane is a greenhouse gas that traps four times more heat than carbon dioxide, its widespread release could push climate change to levels that would be dangerous for human civilization quite rapidly. Under this scenario, further warming of the climate could release the vast deposits of methane trapped beneath the ocean surface in methane hydrates. These are large chunks of ice with methane bubbles trapped inside them. If these reserves of arctic methane are released into the atmosphere, the survival of billions of human beings would be threatened (Romm 2016, pp. 80–82).

To prevent or at least postpone such dire scenarios, many have proposed deliberate efforts to change the earth's climate, which has come to be known as geoengineering. Any attempt at geoengineering initiated by one or several nation states acting within the current Westphalian paradigm of national sovereignty is likely to lead to bitter international disputes, if not war. Considering the prospect of geoengineering, Spencer Weart has raised the obvious question, "What nation would be trusted to alter other nations' weather?" (2008, p. 88). As the historian James T. Fleming has documented, human schemes to alter the climate or "control" the weather have a checkered history that does not inspire confidence (2010).

This is not to say that geoengineering is a hopeless solution, but it is one that should be approached with extreme caution, and with a sober awareness of what Robert K. Merton called the law of unintended consequences (Merton 1976, p. 146). As a sociologist, Merton was aware of the immense complexity of human societies and the inevitable fact that any purposive social action would involve multiple consequences outside of its intended goal. If we consider that geoengineering combines the complexity of human society with the complexity of the earth's climate system and biosphere, Merton's insight about unintended consequences should be at the forefront of every discussion about picking up the tools of geoengineering in an attempt to control a climate system that we have already altered inadvertently through widespread deforestation and the emission of greenhouse gases. If human beings ultimately embrace geoengineering, they would be wise to first create a supranational democracy that could thoroughly debate and carefully monitor any implementation of such policies.

Here is an analogy that might illustrate the difference between possessing the agency to alter a complex system, and possessing control of that system: Imagine that three monkeys find a two-ton pickup truck with its engine idling at the top of a very high hill with a long and winding highway to the valley down below. These monkeys are curious enough to admire the truck, and they explore the truck together until they discover how to climb inside. Each primate finds a part that appeals to his fancy: the first monkey stands behind the steering wheel, the second monkey plays with pedals down below, and the third monkey makes himself at home in the flatbed behind his cohorts. If these primates are clever enough to shift the vehicle into drive, they will soon find that the truck is moving so fast down the hill that none of them dares to jump out. Their fates are tied together with the fate of the careening truck. In this scenario, each of them has a role to play. The first monkey can alter the direction of the truck with the steering wheel but needs the second monkey to hit the brakes or gas in response to the traffic on the highway. Likewise, the second monkey needs the third monkey to tell him when it is not safe to brake because a vehicle is approaching from behind. If, by some miracle, the three monkeys manage to steer the truck to a safe spot on the shoulder of the road and bring it to a stop, they will have achieved the optimum plausible outcome for this scenario. Even if they achieve this outcome, it would be absurd for anyone to claim that these monkeys had learned how to drive a truck. At the most, we could say that they had learned how to cooperate effectively in response to a crisis. In a similar way, the advent of the Anthropocene does not mean that human beings are any closer to controlling nature. Our future actions will greatly impact the fate of the planet, but our actual level of control in this situation is comparable to that of the three monkeys in the careening pickup truck. The best we can do is to learn to cooperate in order to avoid a devastating crash.

An individual human may protest this analogy with the claim that she or he is wiser and more responsible than the analogy of three monkeys in a truck would suggest. On an individual basis, such a claim is usually justified. When we consider the aggregate behavior and impact of our species, however, it is not. The enduring competition between rival nation states is one of the reasons that the collective behavior of our species is less intelligent than it could be. Over the past two centuries, the most powerful social construct for directing the collective activities of human beings has probably been the nation state. Furthermore, competition between nation states has produced the most stunning examples of collective achievements for good or ill, from the development of nuclear weapons to the moon landing and the creation of the Internet. It is therefore not surprising that the intense conflict and competition among nation states that characterized the twentieth century has been a key factor in transforming our relationship to nature. Because of the powerful new technologies that this competition has spurred us to develop, our species has reshaped the earth itself. However, as long as our dogmatic adherence to national sovereignty precludes any effective system of law and accountability, our long-term prospects for survival will hardly be better than those of our simian friends speeding down the mountain highway in a pick-up truck.

Human beings have been doing science, in one form or another, for as long as we have been using language or fashioning tools. The first person to tell another, "Don't eat those berries, they'll make you really sick!" was passing along knowledge about nature that had been acquired through experimentation and, no doubt, confirmed by peer review. Though we live in an era when nation states and corporations fund most scientific research, the human practice of science is older than any of these institutions, just as it is older than numerous kingdoms and empires that once thrived or fell because of its discoveries. When the Nazis condemned the theories of Albert Einstein as "Jewish physics" and anathema to "German science," they revealed a mentality that was not only devoid of humanity, but also of common sense. The sun and moon and stars do not belong to any political or religious entity, and neither can the sciences by which we understand them. Once confirmed, our suppositions about the universe are universal, and they cease to be our own.

All of this should be obvious, but in an age when discussions of science are framed by competition among nation states and corporations, it is not. The twentieth century, which ushered in the age of total war and weapons of mass destruction in which we still live, distorted the human practice of science and irrevocably altered the relationship between humanity and nature. When environmental scholars discuss the Anthropocene, they are liable to disagree about when it began. Some argue that human beings began to have a measurable impact on the earth with the advent of agriculture roughly ten thousand years ago (Ruddiman 2005), while the greater number of earth scientists and environmental historians continue to maintain that the first significant changes began during the early stages of the Industrial Revolution in the second half of the eighteenth century (Davies 2016a, b). There is less disagreement, however, about when human beings became *aware* of their power to transform this planet. This was a revelation that came upon our species over the course of the past century. As the historian of science Rosalind Williams has noted in *The Triumph of Human Empire*, a few cultural luminaries such as Jules Verne had foreseen this change in the late nineteenth century (2013). This was an insight that was largely inspired by accelerating advances in military technology. The idea that modern warfare could reach such a scale and intensity that it threatened to transform – or even destroy – the earth itself had been intimated by the American writer and historian Henry Adams in the second year of the American Civil War. Appalled by the destructive forces that the war had unleashed, he wrote to his brother Charles Francis Adams that "Someday science may have the existence of mankind in its power, and the human race commit suicide, by blowing up the world" (Ferkiss 1993, p. 95).

During the twentieth century, a critical mass of scientists and intellectuals began to describe a fundamental shift in our relationship to the earth. Population increases and rapid advances in mass production, transportation, and communication certainly contributed to the growing sense that humanity had come to have an impact on the natural world that few could have imagined two centuries ago. On the eve of the Great War, the pioneer of science fiction H.G. Wells elaborated on such predictions by depicting the aerial delivery of atomic bombs in his novel *The World*

*Set Free* (1914). Wells concluded his novel with the construction of a postwar utopia that exploited atomic energy for peaceful ends because the human race was now wise enough to embrace the author's favored principles of science, world government, and socialism (Wells 1914). More than any other factor, it was the advent of total war between sovereign nation states, with all of its attendant destruction and technological innovation, which convinced many that our species had now attained the power to transform nature in fundamental ways. The fact that a key marker of our new geological epoch is the global diffusion of radioactive isotopes from nuclear weapons tests illustrates this point. We entered the Anthropocene during a period of total war and sustained strategic competition among sovereign nation states. Nationalism gave birth "the end of nature."

Even before the advent of nuclear weapons, World War One provided the first dramatic example of how military competition and conflict could transform the relationship between humanity and nature. The grisly spectacle of industrial warfare that erupted in August of 1914 was the culmination of a process that had been building for over a century. The exigencies of strategic competition that accelerated the expansion of colonialism throughout the nineteenth century had also fueled intense industrial innovation within Europe. While competition in the field of industry was primarily economic, it always had a military component. This element became especially apparent in the early twentieth century as economic competition between Britain and Germany in the realm of industrial production fueled a more dangerous kind of competition in the form of a naval arms race. Britain had the largest and most powerful navy in the world at the close of the nineteenth century, but Germany, under the rule of Kaiser Wilhelm II, was determined to rival Britain's success in order to protect his own national aspirations for "a place in the sun," or, in other words, a larger overseas empire to provide raw materials, cheap labor, prestige, and markets for German industry (Keylor 2001, p. 8).

Even as the Anglo-German arms race accelerated in the early twentieth century, some economists on the European continent and in Britain attempted to alert anyone who would listen to the economic futility of extending industrial competition into the realm of warfare. The industrialist Jan Bloch had spent decades studying the economic implications of modern war for industrial economies, and he made a number of predictions about what a European war would look like in the twentieth century that proved to be stunningly accurate. Bloch argued the advances in munitions, industrial production, and resupply meant that any future land war in Europe would soon devolve into a costly stalemate leading to the probable economic collapse of the belligerent powers, including the "victors" (Ela Bauer 2016, p. 79). As a Polish Jew living the Russian Empire, Bloch may not have expected his ideas to have the influence that they did. However, Tsar Nicholas II called for a Peace Conference at the Hague in the Netherlands in 1899, partly in response to Bloch's treatise, and partly in response to his growing realization the Russia could not compete in the current arms race with rivals such as the Austro-Hungarian Empire (Blom 2008. p. 194). Although this peace conference yielded little in the way of real reform, it did put Bloch's conclusions about the absurdity of war in the industrial age before an international audience. Some prominent advocates for peace and international law, such as the American industrialist Andrew Carnegie, supported Bloch's ideas and helped to create a Museum of War and Peace in Lucerne, Switzerland in 1902 in order to promote them. However, Bloch's ideas found little favor among political and military leaders across Europe. Needless to say, Bloch's message failed to change the direction of European politics, and the war that he had been unable to prevent proved to be just as calamitous as he had predicted it would be.

Another Cassandra in the years before World War One was the British journalist and author Norman Angell. In 1909, Angell published a pamphlet entitled *Europe's Optical Illusion* which was soon republished through several editions as *The Great Illusion*. An opponent of militarism at home and abroad, Angell argued that the premise that any nation with a modern economy could expand its wealth by annexing territory from its neighbors had been rendered obsolete by modern systems of banking and credit, and by the relentless integration of economic activity across national borders. As Angell put it: "For a modern nation to add to its territory no more adds to the wealth of the people of such nation than it would add to the wealth of Londoners if the City of London were to annex the county of Hertford" (1913, x–xi). Angell's arguments, like those of Bloch, were not heeded, and his attempts to convince the governments of both Britain and Germany to avoid war failed. Unlike Bloch, who died before the First World War, Angell lived to see his warnings borne out, especially concerning the huge price that Britain would pay in the Great War and its aftermath.

In recalling Angell's legacy, it is worth noting that his actual words and actions do not confirm the caricature presented by his critics. The political scientist Daniel H. Deudney observes that Angell "has served for generations as a favorite whipping boy of the Realists" (2007, p. 205). In such critiques by advocates of *Realpolitik*, Angell is depicted variously as a naïve progressive who heralded that wars would no longer happen in the modern age, or as an absolute pacifist who ignored the fact that waging war is sometimes a grave necessity. Neither assessment of Angell's thinking and activism is accurate. He argued before August of 1914 that warfare between modern nation states was irrational and unprofitable, but not *impossible*. Angell's activism prior to World War One furnishes sufficient evidence that he viewed another European war as a danger to be avoided. For years before the Second World War, Angell was not a pacifist, but rather a critic of appeasement and an advocate of military preparedness in the face of the threat posed by the Third Reich (Ceadal 2009, p. 99). In fact, as Deudney observes, it "is doubly ironic that the 'idealism' label sticks so tightly to Angell's rationalist and materialist arguments about war, since his main foil was the idealist conception of the state" that fueled nationalism in both Germany and Britain before the First World War (2007, p. 205).

In the twentieth century, as in earlier times, prophets did not find honor in their own countries. While liberal internationalists such as Bloch and Angell found an audience among citizens of diverse nations across the world, they had little impact on the leaders whose decisions in the summer of 1914 led to the most catastrophic war that history had ever seen at that point. It would, of course, be ridiculous to judge all leaders equally for blundering into the First World War, or any other war for that matter. A state of peace among sovereign nations must always entail some collective effort if it is to last, but the leadership of a single nation, acting alone or with a few confederates, is capable of plunging an entire region into war. The prevention of wars in the future requires the strengthening of those norms and institutions that make it possible for non-belligerent nations to contain the designs of any leader or group that seeks to advance its interests through offensive warfare. Bloch and Angell both supported the creation of such norms and institutions, but the world has yet to see their creation on the scale necessary to prevent the sort of arms race that Bloch identified as absurd more than a century ago. Although the world has succeeded in avoiding another global conflagration for over seven decades since 1945, the major powers of the world continue to arm and prepare for total war. Such a war, if it were to happen by either accident or malevolent design, would utterly destroy the complex web of relationships on which we depend for food, air, and water, and which we commonly refer to as "nature." In an age of total war, competition among nation states has irrevocably altered our connection to nature and distorted the practice of science.

The concepts of total war and the transformation of nature have a deep historical affinity. One of the first individuals to envision the complete transformation of nature as a plausible goal for human beings was Sir Francis Bacon, who argued in his 1620 treatise on the philosophy of science, the *Novum Organum*, that it would be both possible and desirable through science and engineering "to establish and extend the power and dominion of the human race itself over the universe" (Bacon [1620], 2000. p. 100). One of the first individuals to expound the concept of total war was the German general Erich Ludendorff in his book, *Der Totale Krieg* (1935). Ludendorff argued that modern warfare demanded the mobilization of the entire economy and population, and that periods of peace between nations should merely be regarded as intervals between their return to the inevitable and imperative state of war (Heuser 2010, p. 194). On the face of it, the absolutism of both of these concepts renders them paradoxical, and each idea conjures the image of a snake eating its own tail. Regarding Francis Bacon's vision, one might ask how human beings can ever attain a complete dominion over nature if they are in fact part of nature, and subject to its laws. Ludendorff's contention that modern warfare must be total and unending appears equally absurd. Wars are fought for objectives, such as the acquisition of resources or the preservation of a given political system, and these are goals that people expect, once they have "won" the war, to enjoy in peace.

The innate absurdity of an idea is no obstacle, however, to its widespread acceptance. The growing influence of both of these ideas in the twentieth century illustrates this point. Ludendorff's vision of total war had its roots in the full mobilization of the German economy, under his effective command, during the First World War, and it proved to be a concept that both the Axis and the Allied powers would embrace during the Second World War. The U.S. Army Air Force General Curtis Le May employed the concept of total war in his saturation bombing campaigns against Japan from 1943 to 1945, and it would also inform his thinking as he worked to create the Strategic Air Command during the Cold War (Buckley 2006, p. 193). Likewise, Bacon's idea that we could completely transform nature to suit our own purposes would find a growing number of adherents in the twentieth century. As if locked in a runaway feedback loop, febrile visions of total war and the transformation of nature each gave new force to the other during the First World War, the Second World War, and the Cold War. The exigencies of total war spurred scientific research and technological innovation, and the new powers yielded by such research and innovation seemed to hold out the promise that we might dominate nature itself. The war to conquer nature and the war to conquer "the enemy" marched together in the twentieth century. For instance, the pioneering German chemist Fritz Haber first achieved fame when he discovered a method for fixing nitrogen from the atmosphere, thus spurring the creation of artificial fertilizers that revolutionized agriculture in the twentieth century. However, his devotion to Germany also led him to bring gas warfare to the battlefields of Europe in the spring of 1915, sparking a global chemical arms race during the First World War (McNeill 2001, p. 24). The Second World War would continue this pattern on a vastly larger scale. Technologies developed during the struggle between the Allied and Axis powers, such as DDT and the atomic bomb, would be hailed for their ability to eliminate disease and provide energy "too cheap to meter" in the years following World War Two (Pfau 1984, p.187).

By the early 1960s, a number of prominent scientists and intellectuals had come to lament this combination of militarism and a technocratic desire to control nature. What would come to be called the environmentalist movement began to gather force with the publication of Rachel Carson's *Silent Spring* in 1962. Carson traced the creation of DDT and other chlorinated hydrocarbons to the chemical arms race that had begun during the First World War, and compared their lingering effects to the invisible but deadly hazard of nuclear fallout that had become a new source of anxiety during the Cold War. Carson explicitly linked the dangerous overuse of these compounds to the old dream of transforming nature for our own purposes (Carson, R. [1962]. 2002, p. 297). Published shortly before the Cuban Missile Crisis, *Silent Spring* spoke to concerns about war, technology, and the transformation of nature that would soon command the attention of millions.

The power of war to upend our relationship with nature had been much less apparent prior to the First World War. In the late summer of 1914, few had imagined the shocking transformation of the European landscapes that would become the most visible consequence of World War One. In fact, the culture of romantic nationalism that pervaded the first wave of mobilization inspired many writers and propagandists to see the war as a return to nature rather as a catalyst for the transformation of nature. The most famous patriotic English poem of 1914, Rupert Brooke's sentimental sonnet "A Soldier," commandeers the pastoral imagery of Wordsworth in the service of the war effort. In a similar vein, propaganda and advertising from this early period often struck a pastoral note. A 1915 poster published by the British Parliamentary Recruiting Committee featured bucolic images of the British countryside to bolster appeal to prospective recruits: "Your Country's Call – Isn't this worth fighting for?"

Within the first year of fighting, however, the tremendous power brought to bear by machine gun fire, artillery, and the constant resupply of men and materiel via rail networks would create a new kind of terrain on the Western Front. This was the lifeless lunar landscape that the British war correspondent Ernest Swinton labeled "No Man's Land" (Foley 2014, p. 12.). While searing images of No Man's Land were hardly suitable for wartime propaganda, this new and distinctly anti-pastoral landscape was vividly described by the trench poets, such as Wilfred Owen and Siegfried Sassoon who sought to convey the horrors of the Great War. The artist Paul Nash, who first saw the Western Front in February of 1917, would create one of the most iconic renderings of No Man's Land in his 1918 painting, "We Are Making a New World." Echoing the most grandiose propaganda of the period, the irony of Nash's title speaks for itself. In a letter posted to his wife in November of 1917, the twenty-eight year old painter bitterly declared: "I am no longer an artist interested and curious, I am a messenger who will bring back word from the men who are fighting to those who want the war to go on forever. Feeble, inarticulate, will be my message, but it will have a bitter truth, and may it burn their lousy souls" (1949, p. 211).

For some, the tremendous power unleashed by the First World War strengthened their faith in the scientific conquest of nature, even as they surveyed the horrific landscape of No Man's Land. The Scottish biochemist and geneticist J.B.S. Haldane was one of those rare individuals who would look at the deadly lunar vistas of the Western Front and see the birth pangs of a new world blessed by rational government and endless technological progress. After serving as a second lieutenant in the Black Watch Regiment in France, Haldane recounted a little bit of his combat experience in an address to a radical salon called the Cambridge Heretics Society in 1923. Haldane's iconoclastic and provocative lecture was published the following year as a pamphlet entitled *Daedalus, or Science and the Future*. This vision of nature transformed by technology created a stir among British intellectuals in the 1920s, and would be cited decades later by such authors as Freeman Dyson and Arthur C. Clarke as a seminal influence on their own thinking (Dyson 2008. p. 8; Clarke 2001, p. 497). The essay begins with Haldane's recollection of a battle that he witnessed on the Western Front in 1915.

He recalled the artillery and heavy machinery "tearing up the surface of the earth and disintegrating the works of man with an almost visible hatred." From the vantage point of the Western Front, Haldane observed that it was "hard to believe" that the "irrelevant looking human figures" were in fact "the protagonists in the battle" and wondered whether, on a battlefield dominated by such machines "men are in reality their servants, and playing an inglorious, subordinate, and fatal part in the combat". After acknowledging the possibility that advances in science and technology could lead to the enslavement of the human race by machines or even planetary destruction from "a too successful experiment in induced radioactivity," the essay shifts into a surprising tone of resolute optimism. Haldane dismissed the protests against technological progress by contemporary critics such as G.K. Chesterton as not only misguided but irrelevant. Although he regarded himself as a Marxist advocate of international socialism, Haldane saw in both capitalism and nationalism the guarantors of continued scientific research for a long time to come: "Capitalism, though it may not always give the scientific worker a living wage, will always protect him, as being one of the geese which produce golden eggs for its table. And competitive nationalism, even if war is wholly or largely prevented, will hardly forego the national advantages accruing from scientific research" (1924).

Building on his confidence that the accelerating advance of scientific research would continue indefinitely, Haldane also expressed his conviction that the new technologies it engendered would produce radical benefits that would ultimately extend to the entire human race. In a narrative that he characterized as more modest than the predictions of H.G. Wells, Haldane made predictions that were stunningly optimistic: war, imperialism, and economic exploitation would ultimately wither away, as the global integration of industry would render both "international injustice" and "industrial injustice" self-defeating. For the world's energy needs, finite supplies of oil and coal would be replaced by harnessing the wind and sun, once an adequate system for the storage and distribution of energy had been achieved. The problems of soil depletion and world hunger would be solved by advances in applied biology leading to the creation of new strains of wheat and a method for fixing nitrogen in the soil through the use of a new strain of purple algae created under laboratory conditions. To this prediction, Haldane added a grace note to signal just how different the world would be: a synthetic nitrogen-fixing algae, accidentally released into the sea, would spread throughout all of the world's oceans, coloring them a brilliant purple (1924).

Haldane also predicted a complete transformation of human sexuality and family life. In his vision of the future, which he hoped would come to pass before the end of the twentieth century, over seventy percent of human beings on earth would be conceived and gestated in laboratories through the process of ectogenesis. For the life of the individual, one benefit of this change would be to liberate human sexuality from the task of biological reproduction. For society at large, the benefit would be an advance in the practice of eugenics – a cause favored by Haldane and many other intellectuals on both the left and right, especially during the interwar decades (Overy 2009, p. 95–135). Haldane may have been the first biologist to advance the concept of human ectogenesis, and his advocacy of the idea would have a great influence in the English-speaking world, inspiring its satirical depiction in two Aldous Huxley novels, *Crome Yellow* in 1922 and *Brave New World* a decade later.

The optimistic vision that Haldane sketched in "Daedalus" was strikingly original in many respects, but it was hardly unique. The rational organization of industry pioneered by the American Frederick Winslow Taylor had been accelerated by exigencies of the First World War, and many had come to believe that the values of precision and efficiency that characterized science and engineering could be applied universally to economic and social problems. Leftist economists such as Thorstein Veblen argued that "a soviet of engineers" could produce a more efficient and humane pricing system than the marketplace, while the term "technocracy," coined by the engineer William Henry Smyth in 1919, became the nucleus of what promised to be a broadly appealing, if somewhat inchoate and inarticulate, alternative to both the liberal capitalism of the West and the Bolshevik dictatorship that had emerged in 1917 from the ruins of the old Russian Empire (Akin 1977).

In light of the accelerated cultural and political change that followed World War One, it is not surprising that variations on the techno-utopian themes expressed in Haldane's "Daedalus" could be found in a wide variety of places. Haldane's friend and colleague, J.D. Bernal presented his own ideas about the technological transformation of nature in his 1929 treatise *The World, the Flesh, and the Devil*. The Russian biochemist Vladimir Vernadsky expressed a vision similar to Haldane's when he foresaw the evolution of the noösphere – a planetary realm of consciousness that would become integrated with the biosphere through advances in communication technology (Margulis and Sagan 1995, p. 170). The editor and publisher Hugo Gernsback had been fascinated with the idea of a technocratic world state at least since 1911, when he depicted such a world in his novel *Ralph 124C41+*. Gernsback's success in finding readers for this kind of fantasy exploded in the decade following World War One, when he created the pulp science fiction genre with his immensely successful magazine *Amazing Stories*. Among the early contributors to the pages of *Amazing Stories* were two of Haldane's friends, H.G. Wells and Aldous Huxley's elder brother, the biologist Julian Huxley. Among technocratic idealists, Julian had few rivals. A technological optimist whose fervor could rival that of J. B. S. Haldane, Julian Huxley could even see a silver lining in the mushroom clouds that changed the world in 1945. In an address given at Madison Square Garden just two months after Hiroshima and Nagasaki, he sought to persuade his audience of the potential virtues of "atomic dynamite" for building canals and harbors, and even for eliminating the polar ice cap and permanently warming the earth's climate (Deese 2015, p. 118). The fact that Julian Huxley was a pioneering conservationist, who later helped to found the International Union for the Conservation of Nature (IUCN) and the World Wildlife Fund (WWF), would seem to be incompatible with his blithe optimism about a new technology such as the atomic bomb. However, Julian's thinking illustrates the technocratic thinking that was common among many twentieth century intellectuals, including conservationists such as himself. Like Haldane and Bernal, Julian Huxley had immense faith in the ability of scientific experts to lead humanity forward. Although Julian Huxley's proposal to warm the earth's climate with atomic bombs seems risible today, it deserves to be remembered for two reasons. For one thing, it underlines the affinity between military technology and the dream of transforming nature. For another, it should serve as a caution to contemporary advocates of ambitious geoengineering schemes that their proposals may one day look just as absurd.

In the 1920s, outspoken technophiles such as J.B.S. Haldane and Julian Huxley were well accustomed to fielding opposition to their technocratic ideas from religious conservatives such as G.K. Chesterton or romantic iconoclasts such as D.H. Lawrence. However, the ideas set forth in *Daedalus* also inspired an extended rebuttal from an individual who shared Haldane's passion for science. Bertrand Russell, the English mathematician and rationalist philosopher who had been jailed for his anti-war activities during World War One, took such a strong exception to Haldane's predictions that he published a direct rebuttal, entitled *Icarus, or the Future of Science*. As Russell saw it, the fundamental premise of Haldane's optimism about the future of science had scant real evidence to support it. Haldane had assumed at the outset of *Daedalus* that advances in science and technology would somehow be the catalyst for historical changes that would make exploitative capitalism and predatory militarism obsolete. Russell, employing a brutal economic logic that a Marxist such as Haldane should have appreciated, argued that the private businesses and national governments funding scientific research now and in the future would exercise a controlling interest over the very nature and application of that research. In light of this fact, it was extremely likely that scientific and technological progress would have an effect that was precisely the opposite of Haldane's utopian scenario. Barring some fundamental change in the economic and political structure of modern societies across the globe, every new advance in science and technology would most likely deliver the goods that it was paid to deliver: more wealth to the already wealthy, and more power to the already powerful.

Russell predicted that, as a result of this economic and political stratification, science itself would become increasingly distorted by its deepening relationship with private capital and military power. Competition among private corporations and sovereign nation states was not destined to advance science, but to distorts its methods and pervert its application. Although scientific research had begun in the 17th and 18th centuries as an international community in which curious individuals exchanged questions and discoveries with little regard for national borders, the military and economic competition of the twentieth century would regiment science, leading to an increased level of secrecy between scientists around the world, and a loss of autonomy for scientists themselves. In his 1931 book *The Scientific Outlook*, Russell developed the prognosis that he had first advanced in *Icarus* into a broader argument about the changing nature of science. Here he sketched a grim vision of a scientific dictatorship that would apply discoveries in biology and psychology to the efficient manipulation and control of humanity. Russell closed *The Scientific Outlook* with these elegiac remarks on what he believed had been lost:

> Science in its beginnings was due to men who were in love with the world. They perceived the beauty of the stars and the sea, of the winds and the mountains. . .. [A]s science has developed, the impulse of love which gave it birth has been increasingly thwarted, while the impulse of power, which was at first a mere camp-follower, has gradually usurped command in virtue of its unforeseen success (2001, p. 216).

Here Russell sounded an alarm that had a special resonance for some contemporary scientists, as well as artists and intellectuals who sought to integrate the sciences and humanities. In an age of total war and constant strategic among nation states, science was now valued above all as a source of economic, military, and political power.

Russell's vision of technocracy gone awry would in fact form the outline for Aldous Huxley's 1932 novel *Brave New World.* Aldous Huxley had lost most of his Oxford classmates in the First World War, and he was keenly aware that, were it not for his badly damaged eyesight, he probably would have been cannon fodder as well. He was also prescient in his early intimation that the conflict which had begun in 1914 would not be "the war to end all wars" but was likely to be something more akin to the opening conflict in a geopolitical struggle akin to the Thirty Years' War that had torn Europe apart in the early seventeenth century. Like Haldane, Aldous Huxley foresaw that total war would transform our relationship to nature, but, like Russell, he believed that the consequences of this transformation would probably be horrific. Writing in the spring and summer of 1931, Aldous Huxley began his novel *Brave New World* with a cinematic montage of descriptions to give the reader a sufficient glimpse of the devastating "Nine Years' War" that made the construction of the World State with its programmed humanity and omnipotent inquisitors a dire necessity. Modeled on Russell's cautionary vision of a scientific dictatorship, the World State employs science merely as a means to power and has abandoned the pursuit of knowledge for its own sake. Pure science as a form of inquiry is explicitly banished from the World State – only those technologies that placate the masses and enhance state power are permitted to flourish (Deese 2015. p. 65). Of course, Huxley's dystopia was set far in the future, in the Year of Our Ford 632. In the twentieth century, it would be the nation state, not the world state, which would drive the most significant research in science and technology.

The impact of nationalism and total war on the pursuit of scientific knowledge in the twentieth century is well illustrated by the story of Leo Szilard. As a physicist who had been drafted into the service of the Austro-Hungarian Empire during the First World War, Szilard had an intuitive sense of the dangers that expanding militarism posed to science and to the spirit of free inquiry. Even as a child, he had instinctively distrusted the propagandistic accounts of the war that he had read in Hungarian newspapers because they defied his own sense of statistical logic that one group of nations could not *always* be in the right, while the other was *always* in the wrong (Szilard 1987, p. 432). As he advanced in his career as a nuclear physicist, Szilard was part of the cosmopolitan community of scientists that thrived in Germany during the Weimar Republic, and he saw the destruction of that community which immediately followed the rise of the Third Reich. He was also among the first to see that the very discoveries about subatomic particles that had excited that international community of physicists in the 1920s could be put into the service of militaristic nationalism in the 1930s.

After reading H.G. Wells' 1914 novel *The World Set Free*, Szilard had an epiphany in 1933 about the possibility of nuclear fission through a chain reaction. Aware that Werner Heisenberg had remained in Germany and would lend his considerable talent to the nuclear research programs of the Third Reich, Szilard spent most of the 1930s telling anyone who would listen in Britain and then in the United States that Germany was likely to develop an atomic bomb. In August of 1939, he convinced Albert Einstein to sign a letter to Franklin Roosevelt, alerting the President to this concern and advising him to begin a similar research program in the United States. After the U.S. entered the Second World War, American research on the development of an atomic bomb would begin in earnest with the Manhattan Project, and Szilard would play a major role in the effort, working directly with Enrico Fermi at the University of Chicago.

Although his insights into nuclear fission and his role in beginning the Manhattan Project made him an essential figure in the creation of the first atomic bomb, Szilard soon realized that he had little or no influence over how and when the product of his pioneering vision would be used. In the summer of 1945, he initiated a petition to President Truman calling for an observed demonstration of the atomic bomb as a last resort before dropping it on a Japanese city. Although 155 scientists and engineers working on the Manhattan Project at Chicago and Oak Ridge, Tennessee signed the petition, General Leslie Groves suppressed the document and succeeded in having it classified until 1963 (Lanouette 2014, p. 1–2). Szilard's experience with military authority during his work on the Manhattan Project illustrated the argument that Bertrand Russell had made in *Icarus* and in *The Scientific Outlook* about the ultimate relationship between power and scientific inquiry. In the years following World War Two, both Szilard and Russell would work together on efforts (which would prove to be just as futile as The Scientist's 1945 Petition) to bring nuclear weapons technology under some form of international control.

The American writer and intellectual Lewis Mumford also bore witness to the power of nationalism and war to transform the promise of science and technology. Mumford's deep engagement with the relationship between science, technology, and nature began during the First World War when he took up a correspondence with the Scottish botanist, city planner, and environmental visionary Patrick Geddes. In his 1915 book *Cities in Evolution*, Geddes had laid out his vision for a technological revolution that would bring not the conquest or control of nature, but a new kind of equilibrium between human beings and nonhuman nature. Geddes classified the fossil-fuel based technologies of the industrial revolution as representing the "paleotechnic" phase of human history, which he hoped to see supplanted by a "neotechnic" phase featuring more verdant cities built to a human scale, decentralized industry, and cleaner sources of energy such as hydroelectric power (1915, p. 53–64). Lewis Mumford, in his early twenties and making a very modest living as a writer and editor in Greenwich Village at the time, was so taken with Geddes' vision that he began a correspondence with him that would last for nearly two decades, though the two men would only meet twice.

For Geddes, the optimistic spirit behind his vision of a "neotechnic" era of greater harmony with nature was dealt a serious blow by the First World War. In 1917, his son Alasdair Geddes was killed by a German shell while serving on the Western Front. Patrick Geddes and his wife Anna Morton Geddes were in India at the time, where she died of typhoid fever before he could find the heart to tell her of their son's death. Emotionally crippled by this double loss, Patrick Geddes was nonetheless determined to continue his work in urban planning and his larger mission of promoting his vision of a neotechnic revolution for the human race. Before his death in 1932, Geddes would provide his consultation on a number of urban planning projects in India, publish a biography of the Indian botanist J.C. Bose, and also design the urban core of Tel Aviv.

In his initial encounter with Lewis Mumford, however, Geddes' emotional desperation at the loss of his son was palpable. More than four decades after their first meeting in New York City in 1923, Mumford would still vividly recall its intensity:

> On the day after his arrival in the basement of the New School, which gave out on the garden, he took me squarely by the shoulders and gazed at me intently. "You are the image of my poor dead lad," he said to me with tears welling in his eyes, "and almost the same age he was when he was killed in France. You must be another son to me, Lewis, and we will get on with our work together." There was both grief and desperation in this appeal: both too violent and too urgent for me to handle (1966, p. 345).

Although Lewis Mumford recalled in this 1966 essay that he found often Geddes' behavior to be overbearing and even somewhat manipulative, his actions during this period indicate a nearly religious devotion to the man and to his vision. In 1925, Lewis Mumford named his first-born son Geddes Mumford, and in the lion's share of his correspondence with Patrick Geddes after 1920 he began his letters with the salutation, "Dear Master."

In his 1934 book *Technics and Civilization*, Mumford sought to expand upon Geddes' paradigm of the paleotechnic and neotechnic phases of industrial civilization. In contrast to contemporary figures such as J.B.S. Haldane and Hugo Gernsback who promoted fantasies in which human beings attained the power to dominate and transform nature, Mumford envisioned a gentler and more balanced vision of the relationship between technology and nature. Holding out the hope that we might leave behind the paleotechnic era of "carboniferous capitalism" with its overwhelming dependence on the extraction and combustion of fossil fuels, Mumford heralded the possibility of a brighter and more verdant neotechnic future for the human race (2010, p. 156). In many ways, this transition from a paleotechnic era based on fossil fuels to a neotechnic era based on renewable sources of energy anticipated the alternative energy movement that has come to gather new urgency as a response to climate change.

By the late 1930s, as the shadow of the Third Reich fell across Europe, Mumford maintained his optimism about the possible future for the human race, but tempered it with stern warnings about the growing dangers of militarism and atavistic nationalism. Because they had passionately supported the cause of anti-fascism throughout the 1930s, Lewis and Sophia Mumford were proud when their son Geddes enlisted immediately in the U.S. Army upon his graduation from Stanford. Unfortunately, in the late summer of 1944 Geddes Mumford would lose his life assaulting the Gothic Line in northern Italy. Suffering the same fate that his teacher had suffered, Lewis Mumford was devastated. After waiting for weeks for the definitive word about their son, Lewis and Sophia finally received the final report from the War Department: "A blue Army stamp said: 'Deceased verified 10/7/44.' That somehow killed the last spark of hope. .." (Mumford 1947, p. 336–337).

Lewis Mumford believed that he had lost his son in a just and necessary struggle against rampant and predatory militarism. He expressed a deep gratitude for his son's life and his brave sacrifice of that life, but the spirit of optimism that he had struggled to preserve for years could not survive the Second World War. The atomic bombings of Hiroshima and Nagasaki convinced Lewis Mumford that a new and insidious kind of power had infected the body and spirit of American culture, and he spent the remaining decades of his life railing against what he called the "megamachine" of postwar technocracy and the "pentagon of power" at its center. In 1964, Mumford would write:

> The inventors of nuclear bombs, space rockets, and computers are the pyramid builders of our own age: psychologically inflated by a similar myth of unqualified power, boasting through their science of their increasing omnipotence, if not omniscience, moved by obsessions and compulsions no less irrational than those of earlier absolute systems: particularly the notion that the system itself must be expanded, at whatever eventual cost to life. (1968, pp. 1–8)

Lewis Mumford's prose retained its wit and its ability to penetrate conventions and appearances throughout his long and extremely productive career. The element that it had lost from the years prior to World War Two, however, was the fierce hope that Patrick Geddes had sought to pass, like a torch he was no longer steady enough to carry, to his young disciple in 1920. Total war and the transformation of nature may be chimerical abstractions at their core, but expanding militarism and the transformation of science became verifiable realities between 1914 and 1945, and these had the power to blight the hopes of more than one generation. The idea that Mumford and Geddes had cherished was a vision of science and technology inspired by a sense of balance between nature and human communities. Scientific research yoked to the service of international competition and national security could produce weapons of extraordinary destructive power, but it held little promise for advancing the neotechnic phase of history that Geddes prophesied in the second decade of the twentieth century (Geddes 1915, p. 72).

In the years immediately following World War Two, Mumford participated in the burgeoning world federalist movement, until the heightened tensions of the Korean War drove it far into the margins of American political discourse (Rosenboim 2017. p. 237). In the last decades of his life, he held out little hope for the dream of a neotechnic revolution that Patrick Geddes had envisioned in the early twentieth century. However, there was another young protégé of Geddes who still a vocal advocate for the creation of an ecologically stable civilization on a global scale (Anderson 2002, p. 174). That was the conservationist Benton Mackaye, founder of the Appalachian Trail. Mackaye's vision was firmly grounded in the principle that science and technology must be liberated from the confining logic of zero-sum competition among sovereign nation states. In the 1920s, Mumford had introduced Mackaye to Geddes. For Mackaye, this encounter with Geddes ignited a revolution in his thinking, and he would expand upon the ideas articulated by Geddes for the rest of his life. As Mackaye saw it, the key to achieving the balance of technology and ecology that Geddes had envisioned was the establishment of world law through democratic federalism. Only then would the power of commerce be regulated by "Global Law" for the good of the planet. Comparing the prospect of a global constitution in the middle of the twentieth century to the establishment of the United States Constitution in the late eighteenth century, Mackaye posed the following questions:

> With respect to commerce we may ask: "Will the United Nations in our lifetime emulate the United States in Washington's?" We can point to Chief Justice Marshall's words: "The power over commerce was one of the primary objects for which the people of the United States adopted their government." Will some future "Marshall" some day say: "The power over commerce was one of the primary means by which the people of the world developed world government."? Benton Mackaye, "Toward Global Law" (Mackaye 1968. p. 103).

Implicit in this speculation was Mackaye's conviction that the establishment of world government could facilitate the neotechnic revolution that he had dreamed of, along with Patrick Geddes and Lewis Mumford, since the 1920s. After the Second World War, in contrast to Mumford, Mackaye still held onto that hope. And this vision of a future civilization that combined a democratic global government with sustainable technology would find support from other visionaries. Although Buckminster Fuller did not employ the same terminology as Mackaye, he also saw that the adoption of alternative energy and the rejection of narrow nationalism as two sides of the same coin. In his 1982 book *Critical Path*, Fuller presented an audacious proposal for an "Ultra-High Voltage World Energy Grid" that would "continually integrate the progressive night-into-day and day-into-night hemispheres of our revolving planet" (Fuller 1982, p. 48–50). In addition to allowing the exchange of uninterrupted transmission of renewable energy across the world's hemispheres, Fuller predicted that such a grid would help human beings evolve beyond not only fossil fuels but also nationalism (Fuller 1982, p. 51).

The divergent reactions of Lewis Mumford and Benton Mackaye to the aftermath of World War Two stand as a reminder that the relationship between war and science during the first half of the twentieth century was complex enough to allow for multiple interpretations. On the one hand, this period saw a militarization of science which produced troubling consequences that would last into the following century. On the other hand, this era of global conflict and technological competition gave birth to new scientific discoveries would deepen our understanding of our impact on the earth, and even advance our understanding of climate change (Weart, 2008. pg. 20). The advent of total war in the twentieth century created military technologies that still threaten to destroy life on earth, but the intense scientific competition that it engendered also led to new discoveries that transformed our understanding of the natural world, and our place within it. The complex interplay of these factors would only intensify during the Cold War decades, shaping the growth of the environmentalist movement that emerged in the second half of the twentieth century.

## Bibliography

Akin, William. 1977. *Technocracy and the American dream: The American technocrat movement, 1900–1941*. Berkeley: University of California Press.

Anderson, Larry. 2002. *Benton MacKaye: Conservationist, Planner, and creator of the Appalachian Trail*. Baltimore: Johns Hopkins University Press.

Angell, Norman. 1913. *The great illusion*. New York: G. P. Putnam's Sons.

Bacon, Francis. 1620. (2000. *The New Organon: Cambridge texts in the history of philosophy*. Cambridge: Cambridge University Press.

Bauer, Ela. 2016. Jan Gottlieb Bloch: Polish Cosmopolitanism vs. Jewish Universalism. In *Cosmopolitanism, nationalism, and the jews of East Central Europe*, ed. Michael L. Miller and Scott Urey. United Kingdom: Routledge.

Buckley, John. 2006. *Air power in the age of total war*. New York: Routledge.

Carrington, Damian. 2016. The Anthropocene epoch: Scientists declare the dawn of human-influenced age. *The Guardian*. https://www.theguardian.com/environment/2016/aug/29/declare-anthropocene-epoch-experts-urge-geological-congress-human-impact-earth

Carson, Rachel 1962. (2002). *Silent spring*. New York: Houghton Mifflin.

Ceadal, Martin. 2009. *Living the great illusion: Sir Norman Angell*, 1872–1967. Oxford: Oxford University Press.

Clarke, Arthur C. 2001. *Greetings, carbon-based bipeds!: Collected essays, 1934–1998*. New York: Macmillan.

Davies, S. 2016a. *Adaptable livelihoods: Coping with food insecurity in the Malian Sahel*. New York: Springer.

Davies, Jeremy. 2016b. *The birth of the anthropocene*. Oakland: University of California Press.

Deese, R.S. 2015. *We are amphibians: Julian and Aldous Huxley on the future of our species*. Oakland: University of California Press.

Deudney, Daniel H. 2007. *Bounding power: Republican security theory from the polis to the global village*. Princeton: Princeton University Press.

Dyson, Freeman. 2008. *The Scientist as rebel*. New York: NYREV, Inc.

Ferkiss, Victor C. 1993. *Nature, technology, and society : cultural roots of the current environmental crisis*. New York: New York University Press.
Fleming, James. 2010. *Fixing the sky: The checkered history of weather and climate control*. New York: Columbia University Press, 129,132,217.
Foley, Michael. 2014. *Rise of the tank: Armoured vehicles and their use in the first world war*. South Yorkshire: Pen and Sword Books, Ltd.
Fuller, Buckminster. 1982. *Critical path*. New York: St. Martin's Griffin.
Geddes, Patrick. 1915. *The city in evolution*. London: Williams and Norgate.
Haldane, J.B.S. 1924. *Daedalus, or science and the future*. London: K. Paul, Trench, Trubner & Co.
Heuser, Beatrice. 2010. *The evolution of strategy: Thinking war from antiquity to the present*. Cambridge: Cambridge University Press.
Kaplan, Temma. 2015. *Democracy: A world history*. Oxford: Oxford University Press.
Keylor, William R. 2001. *The twentieth century world: An international history*. Oxford: Oxford University Press.
Lanouette, William. 2013. *The scientists' petition': A forgotten wartime protest*. Retrieved from:http://www.chino.k12.ca.us/cms/lib8/CA01902308/Centricity/Domain/1456/Atomic%20Scientist%20Petition.pdf. Accessed 23 Dec 2013.
Ludendorff, E. 1936. *The nation at war: By General Ludendorff* [English version of *Der totale Krieg*, 1935]. London: Hutchinson & Company, Limited.
Mackaye, Benton. 1951. "Toward Global Law". First published in *The survey*, Vol. LXXXVII, No. 6. June, 1951). Republished in *From geography to geotechnics* (1968). Champaign: University of Illinois Press.
Margulis, Lynn, and Dorian Sagan. 1995. *What is life?* New York: Simon and Schuster.
McKibben, Bill. 1989. *The End of nature*. New York: Random House.
McNeill, J.R. 2001. *Something new under the sun: An environmental history of the twentieth-century world (the global century series)*. New york: WW Norton & Company.
———. 2010. The environment, environmentalism, and international society in the long 1970s. In *The Shock of the global. The 1970s in perspective*, ed. Niall Ferguson et al. Cambridge: Harvard University Press.
Merton, Robert K. 1976. *Sociological ambivalence and other essays*. New York: Simon and Schuster.
Mumford, Lewis. 1964 (Winter). Authoritarian and democratic technics. *Technology and Culture*. 5 1, 1-8.
———. 1995. A Disciple's Rebellion. In *Lewis Mumford and Patrick Geddes: The correspondence*, ed. Frank G. Novak Jr., 345. London: Routledge.
———. 2010. *Technics and civilization*, 156. Chicago: University of Chicago Press.
Nash, P. 1949. *Outline, an autobiography: And other writings*, 211. London: Faber & Faber.
Overy, Richard. 2009. *The twilight years: The paradox of Britain between the wars*, 93–135. New York: Viking.
Pfau, R. 1984. *No sacrifice too great: The life of Lewis L. Strauss*, 187. United States: University Press of Virginia.
Ruddiman, W.F. 2005. How did humans first alter global climate? *Scientific American* 292 (3): 46–53.
Szilard, L. 1987. *Toward a livable world: Leo Szilard and the crusade for nuclear arms control*. Vol. 3. Cambridge: MIT Press.
Weart, Spencer R. 2008. *The Discovery of Global Warming*, 21. Cambridge: Harvard University Press.
Wells, H.G. 1914. *The World set free: A story of mankind*. London: Macmillan.
Williams, R. 2013. *The triumph of human empire: Verne, Morris, and Stevenson at the end of the world*. Chicago: University of Chicago Press.

# Chapter 3
# Cold War Environmentalism

> *"We travel together, passengers on a little space ship, dependent on its vulnerable reserves of air and soil; all committed for our safety to its security and peace; preserved from annihilation only by the care, the work, and, I will say, the love we give our fragile craft. We cannot maintain it half fortunate, half miserable, half confident, half despairing, half slave—to the ancient enemies of man—half free in a liberation of resources undreamed of until this day. No craft, no crew can travel safely with such vast contradictions. On their resolution depends the survival of us all."*
>
> Adlai Stevenson

**Abstract** A survey of the growth of the environmental movement in the second half of the twentieth century suggests that the spike in scientific research during the Cold War decades was essential to improving our understanding of the Earth and its climate. Although massive investment in earth science research and space exploration during the Cold War inadvertently helped to launch the global environmental movement, the military application of that research also threatened to cause a global catastrophe. A focused and deliberate effort to understand and cope with the threat of climate change, if organized under the aegis of a supranational democracy, could provide a more sustainable foundation for scientific research in this century.

**Keywords** Cold war · David Bradley · Anthropocene · Space race · Cybernetics

In spite of the militarization of science during the twentieth century, it must be conceded that the breakneck technological competition among nations also planted the seeds for important discoveries in the field of earth science. The story of the British defense engineer Guy Stewart Callendar exemplifies this. Tasked by Winston Churchill with devising a way to remove the visual obstruction of persistent fog from British airfields during the Second World War, Callendar helped to devise a solution that eliminated fog through the steady combustion of petroleum. This



R. S. Deese, *Climate Change and the Future of Democracy*, Environmental Challenges and Solutions 5, https://doi.org/10.1007/978-3-319-98307-3_3

operation "consisted of a system of tanks, pipes, and burners surrounding British airfields to deliver petroleum that, when ignited, raised the ambient temperature by several degrees – enough to disperse fog and light the way for aircraft operations" (Fleming 2010, p. 129). The system, which burned up six thousand gallons of fuel during the time it took to land a single aircraft, proved to be effective and was hailed by British Command as effecting "a revolutionary change in the air war" (Fleming 2010, p.159). Callendar's experience with changing local weather conditions through the massive combustion of fossil fuels enhanced his insight into the long-term impact of fossil fuel combustion on a global scale. In the late 1930s, Callendar had affirmed his support for the hypothesis, first suggested by the Swedish scientist Svante Arrhenius at the turn of the century, that the human release of CO2 through the combustion of fossil fuels was likely to increase the insulating properties of the earth's atmosphere and cause global temperatures to rise (Weart 2008, p. 6). During the Cold War, Callendar became such an outspoken advocate of research into anthropogenic climate change through the combustion of fossil fuels that the phenomenon came to be known as the Callendar Effect (Fleming 2010, p. 217).

The terminology of environmental discourse of the mid twentieth century, including such popular metaphors as "the population bomb" and "Spaceship Earth," reflected its connection to the technological competition of the Cold War (Deese 2009). As the historian Jacob Darwin Hamblin has documented, the drive to control nature that had been ignited during the global conflicts of World War One and World War Two was further accelerated by the technological competition and apocalyptic brinksmanship between Washington and Moscow. In fact, military research on deliberately engineering climate catastrophes such as monsoons and droughts during the Cold War anticipated the catastrophic tone of much contemporary discourse on climate change (Hamblin 2013). As Clive Hamilton notes, "The technological hubris of American science in the Cold War was mirrored in the Soviet Union. Competition was intense not only for military supremacy in the 'space race' but also in programmes for weather modification" (Hamilton 2013, p. 138). The twentieth century notion, born of military conflict and strategic competition, that we could find the right set of levers to allow us to control the forces of nature, has taken on a new life in the age of climate change. The ambitious plans for controlling our climate documented by Clive Hamilton include the creation of a global industrial infrastructure to suck carbon from the atmosphere and the creation of an artificially enhanced albedo for the planet by seeding the upper atmosphere with reflective particulate pollution such as sulphur dioxide. Hamilton cautions against the severe dangers inherent in the Promethean solutions offered by advocates of geoengineering. It would be a mistake, however, to assume that advocates of geoengineering are blithe technophiles who believe that we can easily master nature. Many of them view it as the least bad option remaining if our efforts to curb greenhouse gas emissions prove to be too little too late. As the economists Gernot Wagner and Martin Weitzman have observed, "Some of the most serious scientists are looking toward geoengineering as an option – not because they like to, but because it may well be our only hope for avoiding a climate catastrophe" (Wagner and Weitzman 2015. p. 103).

Whatever position one takes on geoengineering, the mere fact that it is now being seriously considered in some quarters underlines the need for more transparent and accountable global governance. If geoengineering is attempted, it will be necessary to have a responsive global authority in place to prevent any single nation from deliberately altering the climate. It is even more imperative that any geoengineering project be thoroughly debated, reviewed, and monitored by a governing entity that is transparent and democratically accountable. We should also remember that the promise of geoengineering, like any human promise, is subject to doubt. This principle is poetically stated in the Gospel of Matthew, when Jesus admonishes his followers against swearing oaths: "Neither shalt thou swear by thy head, because thou canst not make one hair white or black" (Matthew 5:36). This two-thousand-year-old religious text contains a relevant insight for an age in which people are contemplating the prospect of human mastery over a system as vast and complex and the earth's climate. As we lack the power "to make one hair white or black" on the top of our own heads, it seems that we should approach the prospect of geoengineering with an acute sense of humility.

Given the Cold War origin of such schemes to control nature on a global scale, it may be worth remembering the sober advice of one the most respected Cold War strategists on the dangers of this kind of interference with the fundamental processes of nature. Reflecting on the troubled history of the twentieth century, Kennan wrote in the early 1990s that:

> There are two things human beings should never attempt to bring under their control. One is the weather; the other his heredity. No one who deplores the many bones of contention that divide the human community today, with all the attendant wars and other kinds of beastliness with which people conduct these conflicts, could ever wish upon society further ones that would develop if ever men were to venture into these forbidden areas. Nature, left with the control of these processes, may at times seem senseless and capricious in the exercise of her powers. But there is at least the saving grace that human beings cannot blame one another when they deplore the results (Kennan 1993, p. 35).

At the end of his long career as a veteran diplomat and public intellectual, Kennan reasoned that the global political and legal implications of deliberately created weather events would be staggering in scale and would be likely to create unprecedented strife in human affairs. Sadly, this warning already seems antiquated in the technological and political climate of the early twenty-first century. The earth's climate has already been altered by human activity, and new gene-editing technologies such as CRISPR have made it possible to alter the genome of many organisms, including human beings (Kozubek 2016).

Since the impact of our industry and technology has already taken our species into both of the "forbidden areas" that Kennan warned about, it may be that our only rational response is a new kind of containment. While Kennan argued in the late 1940s that the United States should shore up its alliances with other free nations to contain the spread of Stalinist totalitarianism, it is now necessary for advocates of democracy across the world to band together in response to the challenges posed by the Anthropocene. We can no longer prevent the human race from having a profound impact on such fundamental aspects of life on earth as the weather and heredity. What we must do, however, is contain that impact by creating new forms of governance that are transparent, democratically accountable, and global in scale.

In many ways, the environmental movement that emerged during the Cold War was an attempt to foster this kind of democratic accountability around the globe. J.R. McNeill observed that, while the roots of environmentalism extend back centuries, "between 1965 and 1980 these roots somehow absorbed additional nutrients and gave flower to something new, a popular environmentalism that bloomed all over the world to become a significant factor in political life" (McNeill 2010, p. 263–264).While previous environmental movements had been local or national in their aims and focus, this period saw the birth of a new and unprecedented "global-scale environmentalism," addressing "such issues as climate change, ozone depletion, population, overfishing, and so forth: matters that pertain to major parts of the globe if not all of it" (McNeill 2010, p. 263–264). At first glance, the fact that Cold War technologies helped give birth to the global environmental movement seems more than a little bit counterintuitive. In his 1969 bestseller *The Making of a Counter Culture: Reflections on the Technocratic Society and its Youthful Opposition*, Theodore Roszak interpreted the emerging environmental movement, along with other protest movements of the time, as a reaction *against* the pervasive technocracy of the Cold War era. Roszak's casting of environmentalism as anti-technocratic has been reaffirmed by several prominent figures in the movement, such as the naturalist Edward Abbey and Earth First! founder Dave Foreman, who have called for a return to pre-industrial values. This paradigm has also been employed by critics of environmentalism, such as the libertarian writer Ronald Bailey who has criticized the movement as rife with "neo-Luddites" who hate technological progress as much as they hate capitalism (1994, p. 12).

While there is ample evidence that the environmental movement drew much of its passion from popular reactions against the technocratic culture of the Cold War era, it is also true that ambitious technological initiatives of that period—the nuclear arms race, the space race, and the mainframe computer race—exercised a formative and lasting influence on the evolution of the new strain of global environmentalism that would finally capture the attention of mainstream media by the 1970s. The nuclear arms race gave the movement its motivation by shrinking the globe overnight with the scale of its weaponry, altering public conceptions of environmental dangers, and fostering an unprecedented boom in geophysical research. The space race furnished the movement with its most powerful metaphors and imagery, from 'spaceship earth' to countless and powerful photographs of the earth from space, as well as personal accounts of our planet's fragile beauty from astronauts and cosmonauts. Finally, the computer revolution furnished environmentalism with its signature methodology—the analysis of complex systems and the forecasting of long-range trends via computer modeling. In many ways, modern environmentalism was a child of the Cold War.

In his essay, "You and the Atomic Bomb," published in October of 1945, George Orwell first employed the term "Cold War" to describe the strategic rivalry that was then emerging between the remaining great powers after World War Two, especially the United States and the Soviet Union. Orwell correctly predicted that the Soviet

Union would soon break the U.S. monopoly on these weapons. When that happened, the possibility of a total war between the Eastern and Western power blocs would become a suicidal prospect for both sides. Tipping his hat to James Burnham's geopolitical prophecy of 1941, *The Managerial Revolution*, Orwell also predicted that the competing power blocs of the nuclear age would come to share some key similarities in their styles of governance, evolving into something like the managerial state envisioned by Burnham, in which technical elites would come to hold increasing power. In other words, Orwell predicted that the Cold War would be a state of constant low-level conflict between evenly matched technocracies. Technocracy, the vague term for a society managed by the expertise of scientists and engineers, had been championed by the radical economist Thorstein Veblen in the early 1920s and had even enjoyed a brief vogue as a millenarian political movement in the early 1930s (Akin 1977). Now it would graduate from a hopeful dream to an increasingly hopeless reality in the new age of perpetual conflict. Orwell was less afraid that the atomic bomb would usher in the apocalypse and more concerned that it would inaugurate an "epoch as horribly stable as the slave empires of antiquity," by bringing ".. . an end to large-scale wars at the cost of prolonging indefinitely a 'peace that is no peace'" (1945).

Roughly nine months after Orwell's essay, two fission bomb blasts in the South Pacific would lend some credence to his observations about where the world was headed, even if their significance was not fully appreciated at first. The popular press treated the first postwar atomic bomb tests, code named Operation Crossroads and held at Bikini Atoll in the South Pacific in July of 1946, without a great deal of seriousness. A French fashion designer, Louis Reard, named his new bathing suit the Bikini that summer, and the name has stuck ever since. Less than a year after it heralded the end of World War Two, the moral and environmental implications of this new weapon had yet to be explored in depth. It became the goal of a young Army physician, David Bradley, to change the way people thought about the Bomb.

Sworn into the U.S. Army in July of 1945 and commissioned to serve as a medical examiner at Operation Crossroads in 1946, Bradley became the first author to present the danger of the bomb in explicitly environmental terms. Published in 1948, *No Place to Hide* described the two bomb tests at Bikini Atoll in vivid and accessible detail, introducing the public to the effects of bomb radiation on plants, animals, and the marine environment. Reviewing the book for the *New Yorker*, the prominent essayist and outspoken world federalist E.B. White described Bradley's chronicle as, "A sort of diary of contamination, a notebook of the last days of an atoll.. .. His laboratory was a paradise, and the experiment in which he was involved was an experiment in befouling the laboratory itself" (1948).

In a number of ways, Bradley's desire to change public opinion about the dangers of radioactive contamination was similar to that of climate change activists several decades later. Frankly hoping to shock a reading public whom he considered to be too complacent about the bomb, Bradley complained that public understanding of "man's new environment" was distorted by popular fantasy and disfigured by government secrecy. Calling for more transparency and public debate about the environmental effects of atomic weapons, the young physician observed in 1948 that:

> The Bikini tests have never received much attention. The accounts of the actual explosions, however well intended, were liberally seasoned with fantasy and superstition, and the results of the tests have remained in the vaults of military security.
>
> That sort of security is itself a superstition, and a dangerous one. It fosters misconceptions, or what is worse, indifference, and ultimately results in procrastination, half measures, and hysteria.
>
> ... The really great lessons of that experiment... belong to no special group but to all mankind. The atomic era, fortunately or otherwise, is now man's environment, to control or to adapt himself to as he can (Bradley, xii–xiii).

Anticipating the work of Rachel Carson and Barry Commoner in the 1960s and 70s, Bradley here touches on the strongest thread of environmentalism since the dawn of the Cold War, the realization that humans have created environmental dangers that we can't readily perceive with our five senses. In pointing out the dangers of environmental radiation, and arguing that government secrecy compounded those dangers, David Bradley rebels against the anti-democratic assumption that the perils and promise of this new technology were best left to experts in the federal government, the military, and a tightly vetted community of nuclear engineers (1948, p. 165). By calling national security a "superstition," Bradley was heralding the emergence of a new sort of dogma backed by institutional power, which could be as inimical to free inquiry and sound science as religious institutions had been in the past. The secrecy and disinformation surrounding nuclear weapons were particularly galling for this young army physician. For David Bradley, *No Place to Hide* marked the beginning of a career of anti-nuclear activism that would extend into the nuclear freeze movement of the 1980s. When read with the benefit of hindsight, Bradley's words shed light on a pernicious false dichotomy that impeded our progress in addressing environmental problems from radioactive fallout to climate change. This has been the rigid conceptual separation, and consequent competition, between the priorities of national security and ecological sustainability. As disasters for which climate change is the systemic cause will continue to proliferate, it will become necessary for the governments of the world's most powerful industrial nations to transcend old distinctions between national security and the promotion of ecologically sustainable policies in their trade and industry. Due to the "increased frequency of natural disasters and accelerated global warming" environmental threats are expected to "trump traditional military ones" and may "suddenly become the top priority" for governments and security experts (Mastny and Liqun 2014, p. 347). If we are to adequately address this challenge, a steadily increasing portion of defense spending in the U.S. and elsewhere should be shifted to addressing climate change catastrophes such as droughts, floods, and extreme weather events as these have already eclipsed both terrorism and the lingering geopolitical rivalries of the post-Cold War era as the greatest threat to the international community and to the safety and prosperity of civilian populations throughout the world.

A number of Manhattan Project scientists shared David Bradley's concern about nuclear tests and proliferation after World War Two, including the man who had first intuited how a fission chain reaction could be harnessed for the production of atomic bombs. During the Cold War, Leo Szilard continued to feel an acute sense of responsibility for the dangers posed by the nuclear arms race. Known for his feverish creativity, Szilard proposed numerous schemes for the international control of nuclear technology, including the possibility of creating global democratic institutions that would bypass national governments. In 1949, he called for "the creation of a world assembly that would be representative of the peoples of the world, rather than the governments of those people" (Szilard 1987, p. 38). Many of his ideas were dismissed as utopian, but some of them found traction. Szilard's Pugwash Conferences on Science and World Affairs, which first brought together scientists from the Eastern and Western blocs in 1955, would become a lasting global institution. His proposal for a hotline between the Kremlin and White House was eagerly embraced by both the Americans and the Soviets after the Cuban Missile Crisis of October, 1962 (Szilard 1987, xvii).

Szilard's 1961 satirical novella, *The Voice of Dolphins*, may shed light on why he shifted his studies from nuclear physics to biology and environmental issues during the Cold War. Influenced by the dolphin research of his acquaintance John C. Lilly, the iconoclastic polymath sketched a utopian scheme that reimagined his long-cherished notion that an enlightened group of thinkers could transcend nationalism, ideology, and religious dogma and find a path forward for the human race. In this story, however, those enlightened thinkers weren't necessarily human. In *The Voice of the Dolphins* the human race appears to receive sage advice from a group of dolphins living at a research institute in Vienna. Communicating with the scientists who have cracked the code of their language, the brilliant marine mammals offer solutions to a number of pressing global problems, such as producing an adequate food supply and bypassing religious taboos against birth control. At the close of this tale, Szilard leaves it to the reader to decide whether the dolphins had really provided these solutions, or whether the human scientists had used them as a ruse to make their own ideas more palatable to a public that had come to have less trust in human scientists than in these charismatic marine mammals. In the sixties and seventies, the prospect of communicating with dolphins was not only a subject for scientific research, but also had a growing hold on the public imagination. Because he died in 1964, Szilard did not live to see the expansion of this idea in the further, and increasingly unorthodox, research of John C. Lilly, and in countless popular culture tropes, from the television show *Flipper* to the Cold War thriller *Day of the Dolphin*. Although it sometimes reads like a laundry list of proposals that Szilard had already made for arms control and global governance, his novella from the dawn of the 1960s remains a remarkable cultural document. *The Voice of the Dolphins* did more than imagine a *deus ex machina* for the human race in the form of super-intelligent dolphins. The tale is an implicit declaration by one of the century's leading physicists that the human race would need to respect the wisdom of other species before it could find a way out of its own technological labyrinth.

Leona Marshall Libby, another early witness to the dawn of the atomic age, would also turn her attention to environmental issues during the Cold War decades, though from a very different political perspective than either Bradley or Szilard.

Libby, who had been told to pursue more feminine goals when she began her study of physics, earned the respect of Enrico Fermi and in 1942 was the only woman among the handful of scientists who engineered the first controlled nuclear reaction under Stag Field at the University of Chicago. Unlike David Bradley, she remained deeply committed to the idea that nuclear engineers were the most qualified to assess both the potential dangers, and the potential uses, of this new technology. Counting Edward Teller as a close personal friend as well as a personal hero until the end of her life, Leona Libby endorsed the development of the Hydrogen Bomb, under Teller's leadership, and argued passionately with her former mentor Enrico Fermi because of his opposition to the project (1979, pp. 242–244, 302–304). Even the dirtiest spate of atmospheric bomb tests, beginning with the 50 Megaton Soviet Tsar Bomb in October of 1961 and continuing until the Atmospheric Test Ban Treaty of 1963, was viewed by Leona Libby as an opportunity for scientific discovery: "The result is that there was a sharp well-defined injection of radioactivity into the stratosphere in 1962, which has provided us with a global experiment of tremendous importance for understanding the exchange time between various parts of the atmosphere and the oceans" (1979, p. 317). Dividing her time between teaching at UCLA and working for the RAND Corporation in the 1960s and 70s, Leona Marshall Libby wrote several books at RAND that epitomized her technocratic approach to environmental issues.

Leona Libby was hailed by Rainer Berger at UCLA as, "one of the modern pioneers of climatic research" in her early 1980s innovative method which used isotopes and tree rings to measure past climates (1983, ix). While her stalwart faith in both nuclear weapons and civilian nuclear power placed her well beyond the pale of popular environmentalism in the 1960s and 70s, it is worth noting that many of Leona Marshall Libby's views have been embraced by some prominent environmentalists since her death in 1986. The issue of human impact on climate, on which Leona Marshall Libby conducted important research, has proven to be more significant than most of her contemporaries could imagine (1983). When many prominent environmentalists passionately opposed nuclear power in the 1970s, Libby pointed to climate science and argued that nuclear power plants were preferable to the further construction of fossil fuel plants.

Leona Marshall Libby was not alone in seeing a connection between the arms race and earth sciences. The thermonuclear arms race, because it transformed the whole world into a potential battlefield, spurred a wave of government investment in earth sciences. As historian Spencer Weart puts it, "geoscience was one of the privileged fields" because military planners "realized that they needed to understand almost everything about the environments in which they operated, from the ocean depths to the top of the atmosphere. For good practical reasons, then, the US government supported geophysical work in the broadest fashion" (Weart 2008, 21).

In 1957, the earth sciences became a focal point for limited cooperation between the Eastern and Western blocs when both the United States and the Soviet Union joined the majority of industrialized nations in observing the International Geophysical Year (IGY). The IGY led to an impressive array of research and policy breakthroughs, especially regarding Antarctica, but it also inaugurated a new phase of intense Cold War competition when the Soviet Union marked the occasion by launching Sputnik, the earth's first artificial satellite, on October 7 of that year. This was followed by a remarkable succession of Soviet firsts in space exploration, including the first man in space, Yuri Gagarin in April of 1961, and the first woman in space, Valentina Tereshkova, in June of 1963. The U.S. response to these achievements included the creation of the National Aeronautics and Space Administration (NASA) in 1958 to promote space exploration and the Defense Advanced Research Projects Administration (DARPA) that same year to fund new technological initiatives, including DARPAnet, the military and academic precursor to today's Internet.

The story of astronaut Russell Schweickart further illustrates how the space race between the U.S. and Soviet Union helped to catalyze a growing awareness of the earth as a single system during the 1960s. For Russell L. "Rusty" Schweickart, who was just shy of his 22nd birthday and serving in the 101st Tactical Fighter Squadron when Sputnik began its orbit, the successful Soviet space shot was primarily a source of excitement, as were the dramatic successes that followed. In contrast to the public sense of foreboding caused by Sputnik, Schweickart, reports that, "I felt fine about the early Soviet achievements. And frankly all the US astronauts did." Schweickart recalls that he and his colleagues "more interested in seeing humanity get into space than in the political battles. We all knew very well that if the Soviets did something impressive it would further solidify the US commitment to do it even better" (2011).

The U.S. did surpass the Soviets in space when the Apollo program, first conceived during the Eisenhower era and announced by President Kennedy on May 25th, 1961, began to record its first dramatic successes. Schweickart, who served as Lunar Module Pilot (LMP) on Apollo 9, made history by successfully piloting the LMP in earth orbit, and by carrying out one of the first Extra Vehicular Activity (EVA) sessions without a tether or umbilical cable. Schweickart also made history by sharing his personal impressions of seeing the earth from space. Dubbed the "hippie astronaut" by the press for his long hair and poetic descriptions of seeing the earth from space, Rusty Schweickart provided a verbal and personal complement to the famous images of the earth from space. Recalling his experience of conducting an EVA during the Apollo 9 mission, Schweickart departed from the laconic technical jargon that the public had come to associate with test pilots and astronauts:

> I'm just floating there almost as if I'm naked in space. And all, all of this stuff starts coming into my mind: I'm here because life has evolved on this planet, we've developed brains which enable us to invent machines. In combination with those machines we are able to extend our environment and here I am on the frontier of this evolutionary process. What am I? I'm a representative of life moving out into the universe. So, the idea of mother earth, that phrase has real meaning, from the outside you can look back—the child now sees its mother. We human beings, we're this life form on this incredible planet, it's coated with life, where are we going? (Schweickart, *Earth Days* 2009).

The "hippy astronaut" was not alone in pondering these questions. During the 1970s Schweickart served as chief science and technology advisor to California Governor Jerry Brown. He then joined forces with cosmonauts Alexey Leonov, Vitaly Sevastyanov, and Georgi Grechko and founded the Association of Space Explorers

(ASE) in 1984. An international organization open to all individuals who have traveled in space, the ASE currently has 325 members from 35 countries. From the beginning, ecological issues were a primary concern of this new international organization. Schweickart recalls, "During the discussions leading to the formation of the Association of Space Explorers (ASE) all of us in the final planning meeting unanimously agreed to making the theme of our first Congress 'The Home Planet' and selecting our first annual ASE award winner to be Jacques Y. Cousteau" (2011). For his part, Cousteau had been an early and passionate advocate of the space program because he foresaw, quite correctly, that satellite monitoring would greatly increase our understanding of marine ecosystems.

Another astronaut who would turn her attention to environmental issues was Mary L. Cleave. A ten-year-old girl living in Great Neck, New York when the Soviet launch of Sputnik shocked the American public, Cleave later recalled that the adults around her could not help but view the event through a Cold War prism:

> I grew up in Great Neck, NY and so had personal ties to people involved with what was going on in [the temporary UN headquarters at] Lake Success and the early UN activities. Our next-door neighbors on Brown Court were Soviet spies so the cold war was very real for us and early success of Soviet space program was definitely noticed with concern (2001).

Regarding her own personal reaction to Sputnik, Cleave adds, "It definitely increased my interest in the space program" (2011). In college, Cleave would study microbial ecology and go on to publish research on the effects of oil shale leachates on aquatic ecosystems in the American southwest. On her two shuttle missions, in 1985 and 1989, she brought an ecologist's eye to her work in space. Observing the biosphere from Low Earth Orbit (LEO) on two missions 4 years apart, Cleave also gained a heightened awareness of the fragility of the earth's ecosystems and the accelerating pace of environmental change in the late twentieth century:

> I majored in ecology and then environmental engineering because of my concern about the environment and human pressures on it and was really surprised on my first flight.. .. Our mark on the planet was much more obvious than I had expected from LEO and that was probably the most distressing thing to me. In particular, the amount of sedimentation and deforestation was distressing.. .. The view of the very thin layer of atmosphere around the Earth also surprised me. On my first flight, there was a dust storm coming off the west coast of Africa and we watched as the dust cloud moved over the Atlantic Ocean and took about four days to reach Florida. The natural barriers that oceans provide did not seem to have the integrity I had assumed before. On my second space flight, it looked like the issues that concerned me on the first flight were worse, so I started to think about moving back into the environmental business. The importance of the Earth science program at NASA, where planetary science tools are used to study Earth, was very obvious to me after two trips to LEO. . . so, I moved up to GSFC [Goddard Space Flight Center] to work on a project that was planned to measure the plants in the ocean. We needed this information to drive global Carbon models (2011).

Mary Cleave's concern about the ecological changes she could see on her two shuttle missions drove her to seek additional data from satellite instruments to monitor changes in the earth's biosphere.

In the late 1980s, this drive for more satellite monitoring of ecological trends was championed by NASA scientist James Hansen. In the 1960s, Hansen had conducted analysis on the atmosphere of Venus, a planet that the Soviet Union had taken an early lead in researching with its series of Venera probes. Along with other researchers who studies spectroscopic analysis of the Venusian atmosphere, Hansen concluded that the planet's extremely high temperatures were maintained in large part by the heat-trapping gases in its atmosphere. By the early 1970s, Hansen began his own research on whether human emissions of carbon dioxide and other heat-trapping gases were causing average temperatures on earth to rise. The Swedish chemist Svante Arrhenius had predicted this effect in the late nineteenth century, and his hypothesis had gained support from the research of Guy Stewart Callendar and Charles David Keeling. In 1981, Hansen joined these earlier researchers in reaching the conclusion that anthropogenic climate change was indeed a reality, and he published his findings in *Science* that year. In 1988, as head of NASA's Goddard Institute for Space Studies (GISS) at Columbia University, Hansen testified before Congress that global warming was a serious problem. Like Mary Cleave, Hansen argued that greater satellite monitoring of earth's biosphere was essential and threw his support behind NASA's ambitious Mission to Planet Earth (MTPE) in 1989. As part of this ambitious project NASA has launched 22 satellites in the past 20 years to monitor the earth's oceans, atmosphere, and its rapidly shrinking cryosphere. Today this system of satellites returns a steady stream of data as part of the Earth Observing System (EOS).

As Hansen himself recognized in the late 1980s, one of the challenges posed by satellite monitoring of the earth's environment would be finding a way to process the tsunami of data that such systems would deliver in real time. The question of how to understand and predict the behavior of dynamic systems with a huge number of variables had been addressed by military researchers at MIT's Radiation Laboratory (or Rad Lab) during the Second World War. Addressing the challenge of defense against aerial bombardment, these engineers were trying to develop more responsive servomechanisms for anti-aircraft guns. The mathematician Norbert Wiener, who was a lead researcher at Rad Lab, coined the term cybernetics to describe the role that information and feedback played in both biological and artificial systems of command and control, and he published a groundbreaking book on the subject in 1948, *Cybernetics: Control and Communication in the Animal and Machine*. As Wiener's title indicated, Cybernetics was a biomimetic discipline, searching for workable models in biological systems, even as it pioneered new technologies such as computing, remote sensing, and robotics.

While the term cybernetics, and especially the prefix "cyber," have become cultural touchstones in the past several decades, Wiener's pioneering work on the concept of feedback may be even more significant. Anyone who has held a microphone too close to an amplifier has experienced the shock of audio feedback, and anyone who has appreciated the guitar playing of Jimi Hendrix has respect for how feedback can be both explosive and stunningly complex. If we accept the poet Ezra Pound's proposition that "Artists are the antennae of the race," the virtuoso use of feedback in such Hendrix classics as his 1969 rendition of "The Star-Spangled

Banner" resonates in more ways than one. By the middle of the 1970s, "scientists were seeing feedback cycles everywhere, poised to react with hair-trigger sensitivity to external influences" (Weart 2008, p. 74). We cannot understand climate change without considering the many feedback loops in our planet's climate system.

On a more fundamental level, we cannot understand the history of science and technology without recourse to Wiener's concept of feedback. If we accepted a very simple definition of progress, we might depict the history of science and technology as a linear affair in which new scientific discoveries lead to new technologies, which in turn lead to the attainment of stable human goals. If we recognize the role of feedback, however, we can see that the process is anything but linear. Scientific discoveries indeed lead to new technologies, but new technologies also change the nature of science itself, as when the optical and later the radio telescope transformed astronomy. Furthermore, discoveries and inventions destabilize human goals, or even our definition of what it means to be human. Progress, in such a complex system can only be defined as specific progress toward a specific goal, and not as universal progress toward a universally agreed upon set of goals. Though he was indeed a visionary, Wiener could probably not have imagined the multifarious and surprising ways in which his ideas would reverberate in the age of globalization, the Internet, and climate change.

Like Leo Szilard, Norbert Wiener resolutely avoided military research after World War Two. However, the foundation laid by the Rad Lab team proved to have new applications in the age of nuclear weapons. Furthermore, it would have a significant impact on the evolution of Cold War environmentalism. Whirlwind, a computer developed by MIT researcher Jay Wright Forrester to serve as a flight simulator for the Navy in 1944, found a new application after 1949 as the basis of SAGE, or the Semi-Automatic Ground Environment. A more elaborate version of the anti-aircraft systems developed during World War Two, SAGE was designed to coordinate the U.S. air and ground response to a nuclear attack by the Soviet Union. Forrester's Whirlwind was a new breed of computer, employing a magnetic core memory and displaying calculations in real time through a cathode ray tube display. By the late 1950s, Forrester moved to MIT's Sloan School of Management and began applying the principles that he had developed at Rad Lab to management questions. By the late 1960s, his work caught the attention of Italian industrialist Aurelio Peccei, who was looking for a way to forecast global trends in population growth, resource use, and pollution. Peccei invited Forrester to participate in a global think tank he had formed in 1968 called the Club of Rome, which included the world federalist and ocean conservationist Elisabeth Mann Borgese. As with nuclear technology and space exploration, the growth of computer science during the Cold War came to transform our view of the environment and our place within it.

Forrester, along with the biophysicist Donella Meadows and her economist husband Dennis Meadows, headed up a team at MIT which created a computer model of global trends in population growth, GDP growth, resource extraction, and pollution. In 1972, Donella Meadows was the lead author of the group's report, entitled *The Limits to Growth: A Report for the Club of Rome's Project on the Predicament of Mankind*. The book sold over nine million copies in the early 1970s and was translated into 28 languages, but it was also attacked as an exercise in Malthusian doomsaying by many contemporary economists. As the authors of the report were themselves willing to concede in the 1980s and 1990s, their predictions were far from perfect. In a 1982 book entitled *Groping in the Dark,* Donella Meadows at once addressed the difficulty and affirmed the necessity of projects such as the *Limits to Growth* with a Sufi tale about a man looking for a lost key on a sunlit street. When a passerby asks him where he last saw the key, he answers, "I dropped it inside my house. But it's dark in there, so I am looking for it out here." As Donella Meadows acknowledged, attempts to predict future trends through computer modeling were indeed a form of groping in the dark, but they offered greater hope for determining future resource trends than merely parroting the conventional wisdom of the present.

In the years since its publication, *The Limits to Growth* has often been lumped together with Paul Ehrlich's 1968 bestseller *The Population Bomb* as yet another environmental jeremiad proven false by subsequent decades. This comparison is not entirely fair however, since Ehrlich made very specific predictions about the 1970s and 1980s (which have indeed turned out to be false), while the *Limits to Growth* makes probabilistic statements about likely trends across a full century. One of its few near term predictions, that the earth's population in the early 1970s would approximately double by the early 2000s, has turned out to be correct. Otherwise, the general message of *Limits to Growth* is best summed up by Jay Forrester in his precursor study, *World Dynamics*. He declared that the objective of the Club of Rome was to call attention to the finite nature of key resources and "to understand the options available to mankind as societies enter the transition from growth to equilibrium" (Forrester 1971, p. 8). While pro-growth economists such as Julian Simon have often argued that technological innovation will solve the problems of population growth and resource consumption, Jay Forrester reasoned that in the long run such predictions, however optimistic, cannot square unlimited economic growth with limited planetary resources. Although many critics dismissed *Limits to Growth* as overly pessimistic, Forrester sought to present a rational vision of human progress in which a brighter future could be achieved by slowing economic growth while using more efficient technologies to raise the standard of living and the quality of life for all. Forrester's call for an economic focus on quality rather than quantity was similar in some respects to the concept of a neotechnic society envisioned by Patrick Geddes and Lewis Mumford decades earlier. In both cases, the goal of perpetual economic growth was supplanted by a new goal of balance and ecological sustainability.

The tangled relationship between Cold War technologies and the evolution of global environmentalism after 1945 points to a couple of tentative conclusions, and leaves us with at least one unanswered question. The first tentative conclusion is this: Monolithic views of one's ideological rivals can lead to blind spots. Just as a Western belief in monolithic Communism caused many policymakers to ignore the significance of the Sino-Soviet split for nearly a decade, so the monolithic characterizations of a Cold War technocracy on the one hand and environmental rebellion on the other hand have obscured a more nuanced view of both the "technocratic" establishment and the allegedly "luddite" Counter Culture that opposed it. A monolithic view of technologists ignores Leo Szilard's constant clashes with the War Department, even as he brought the Manhattan Project into being, as well as his playful speculation in *The Voice of the Dolphins* that the human race might enhance its chances for survival by learning from other species. Likewise, a monolithic view of the Counter Culture ignores the fact that while many dismissed the U.S. moonshots as Cold War gamesmanship and a waste of money, one of the New Left's most prominent troublemakers, Abbie Hoffman, dismissed such criticism of the Apollo program with characteristic Yippee glee, when he declared: "Are you kidding? We're going to the fucking MOON!" (Brand 2009, p. 214). Most important, a monolithic view of Cold War technocracy on the one hand and counter cultural environmentalism on the other ignores how both trends grew out of the same historical context and came to influence each other through an often messy but sometimes quite creative dialectic.

The second tenuous conclusion we can draw may prove to be quite relevant to addressing the challenge of climate change in this century. It is simply this: ambitious investments in science and technology, regardless of their initial motivation, very often lead to a greater understanding of the natural world. The research generated by the nuclear arms race was purely strategic to begin with, and yet it provided the catalyst for the exponential growth of the earth sciences after 1945 and led to the development of unprecedented tools for understanding the history of the earth's climate. The initial impetus for the space program was international prestige and the more efficient delivery of thermonuclear weapons, but it provided the first photographs and personal accounts of the earth from space, as well as creating the satellite infrastructure for monitoring changes in the earth's biosphere in real time. Likewise, the explosion of mainframe computer technology, first funded to guide anti-aircraft guns in the Second World War and then greatly expanded in the service of the Strategic Air Command during the 1950s, led to unprecedented models for understanding dynamic trends with serious ecological impacts, from population growth, resource extraction and pollution in the 1960s to ozone depletion and climate change in subsequent decades.

Half a century after the feverish technological competition of the space race during the 1960s, we are still faced with this question: Without the bogeyman of a foreign enemy, what could possibly be the next big thing to inspire the sort of massive research and development projects that characterized the Cold War decades? Fear, usually couched in the language of national security, has hitherto been the most reliable motivation for ambitious investment in science and technology research. As a response to the "technological Pearl Harbor" of Sputnik and subsequent Soviet achievements in space, U.S. federal spending on Research and Development skyrocketed from less than one half of one percent of GDP in the mid-1950s to 2% of GDP in 1964. However, in the two decades since the collapse of the Soviet Union in 1991, federal spending on research and development has remained closer to its pre-Sputnik levels (Congressional Budget Office [CBO], 2008). Although the U.S. military build-up since the terrorist attacks of September 11, 2001 has caused some increase in military research and development, there have been no U.S. government projects in the last 40 years that have come close to rivaling the Apollo program in ambition and scale.

While some may hope that recent Chinese achievements in space may create a new space race in this century, it may be wiser to question whether the human instinct to be motivated by fear can be educated to operate beyond the old parameters of international rivalry. In other words, if human societies can be motivated by real or perceived threats to their "national interest" to make massive investments in science and technology, is it possible that we could be inspired to make similar investments to deal with present or emerging threats to the entire planet? Arguing that science and democracy have a symbiotic relationship, Timothy Ferris advances the claim that any nation that invests "at least 2 percent of [its] GDP into scientific research and development" will join the ranks of those advanced nations that enjoy the superior prosperity and quality of life that come from "a growing domestic science and technology sector, free markets, and free government" (Ferris 2010. p. 12).

However, Ferris does not consider what could happen if investment in scientific research and development could be liberated from the paradigm of the nation state. We have already seen how individual nation states can spur science and technology when confronted by a clear and present danger to their national security. Because climate change represents a clear and present danger to global security, a coordinated program of research and development on a global scale is required. Scientific and technical advances in the realms of energy efficiency, clean energy, food production, reforestation, and carbon sequestration could all benefit from a coordinated program of funding on a global scale. If, as Ferris argues, investing 2% of a nation's GDP on scientific research and development will enhance its prosperity, there is reason to believe that a comparable investment of global GDP would produce even greater returns. One of the chief benefits of establishing democratic institutions on a global scale would stem from the powerful investments in cooperative research and development that such institutions would be able to make, drawing from such revenue sources as a universal carbon tax.

Global funding of scientific research for the understanding and preservation of the global commons would free it from the limitations imposed by the logic of the national security state. The militarization of science during the twentieth century eclipsed an older tradition in which scientists shared information across national and cultural borders. Though competition among nation states advanced earth science in the twentieth century, it did this only as a side effect. The main practical legacy of such competition was to create stockpiles of nuclear, chemical, and biological weapons that constitute a continuous threat to the biosphere. The main theoretical legacy of militarized science was to proliferate the idea that human beings might one day be able to completely control nature. Current schemes for the control of climate on a global scale have their clearest antecedents in the militarization of nature envisioned during the conflicts of the twentieth century (Hamblin 2013).

Finally, the third broad conclusion we can draw is certainly the most hopeful and perhaps the most important: the efforts of progressive scientists and intellectuals to cooperate across national borders in the twentieth century led to the creation of new organizations, such as the Pugwash Conferences and the Club of Rome, to assess our prospects for survival and our impact on the planet. If such cooperation could be increased, and safeguarded from the military and economic competition of nation states, it would enhance the prospects of achieving what Patrick Geddes conceived of as the "neotechnic" era for the human race, in which renewable sources of energy would replace our "paleotechnic" dependence on fossil fuels. The best safeguard for this type of cooperation would be the creation of new democratic institutions that would not be under the authority of any single nation state.

## Bibliography


Bailey, Ronald. 1994. *Ecoscam: The false prophets of ecological apo calypse*. New York: St. Martin's Press.

Berger, Ranier. 1983. "Forward" in Libby, Leona Marshall. In *Past climates: Tree thermometers, commodities, and people*. Austin: University of Texas Press.

Bradley, David. 1948. *No place to hide*. Boston: Little, Brown.

Brand, Stewart. 2009. *Whole earth discipline: An ecopragmatist manifesto*. New York: Viking Penguin.

Cleave, Mary L. 2011. *Interview via email.*

Congressional Budget Office (CBO). 2008. Issues and options in infrastructure investment appendix a: Spending for research and development and for education.

Deese, R.S. 2009. The artifact of nature: Spaceship earth and the dawn of global environmentalism. *Endeavour* 33 (2): 70–75.

Ferris, T. 2011. *The science of liberty: Democracy, reason, and the laws of nature*. Harper Perennial.

Forrester, Jay W. 1971. *World dynamics*. Wright-Allen Press.

Hamblin, Jacob Darwin. 2013. *Arming mother nature: The birth of catastrophic environmentalism*. Oxford: Oxford University Press.

Hamilton, Clive. 2013. *Earthmasters: The dawn of the age of climate engineering*. New Haven: Yale University Press.

Kennan, George F. 1993. *Around the cragged hill: A personal and political philosophy*. New York: W. W. Norton.

Kozubek, Jim. 2016. *Modern prometheus: Editing the human genome with Crispr-Cas9*. Cambridge: Cambridge University Press.

Libby, Leona. 1979. *The uranium people*. New York: Crane Russak & Charles Scribner's Sons.

Mastny, Vojtech, and Zhu Liqun. 2014. *The legacy of the cold war: Perspectives on security, cooperation, and conflict*. Plymouth: Lexington Books.

McNeill, J.R. 2010. The environment, environmentalism, and international society in the long 1970s. In *The shock of the global. The 1970s in perspective*, ed. Niall Ferguson et al. Cambridge: Harvard University Press.

Orwell, George. [1945]1968. You and the atomic bomb. In *In front of your nose, 1945–1950*, ed. Sonia Orwell and Ian Angus. London: Secker & Warburg.

Schweickart, Russell L. 2009. Quoted in the film *Earth Days*, directed by Robert Stone. WGBH.

———. 2011. Interview via email.

Szilard, L. 1987. *Toward a livable world: Leo Szilard and the crusade for nuclear arms control*. Vol. 3. Cambridge: MIT Press.

Veblen, T. 1921. *The engineers and the price system*. New york: BW Huebsch. Incorporated.

Wagner, Gernot, and Martin L. Weitzman. 2015. *Climate shock: The economic consequences of a hotter planet*. Princeton: Princeton University Press.

Weart, Spencer R. 2008. *The discovery of global warming*, 21. Cambridge: Harvard University Press.

White, E.B. 1948. *Review of no place to hide. The New Yorker*. December 4th, 1948.

# Chapter 4
# The Tragedy of a False Dichotomy

> *"To those who are awake, there is one world in common, but of those who are asleep, each is withdrawn to a private world of his own."*
>
> Heraclitus

**Abstract** Two false dichotomies have distorted our thinking about environmental issues since the mid-twentieth century. The first is a tendency to place human rights and ecological sustainability in opposition to each other. This tendency was epitomized by those Malthusian environmentalists who advocated both eugenics and strict governmental control over human reproduction in the name of ecological sustainability. The second false dichotomy is the tendency to view technology and nature as antithetical. This has been evident across the spectrum of popular culture, where some of the most acclaimed films with environmental themes frame nature and technology as inherently opposed to one another. Each of these false dichotomies must be transcended if we are to understand and address environmental challenges such as climate change.

**Keywords** Hardin · Tragedy of the commons · Bacon · Technology · Nature

The shape of our world and its place in the cosmos both defy common sense. From the vantage point of our day to day experience, the earth appears to be flat, and the sun appears to rise in the east and set in the west. The knowledge that the earth is round, that it orbits a middling star on the edge of the Milky Way, and that its biosphere is a unique and fragile system that nourishes and sustains us is still news to our species. Our maps of the world are flat, and their most conspicuous element is that patchwork of colors representing what Benedict Anderson called the "imagined communities" of nation states (1983). While most of us know that the sun is not moving in the sky, we speak of it as rising and setting, as though it and all the other stars in the sky still revolve around us. Finally, the biosphere of the planet still seems



R. S. Deese, *Climate Change and the Future of Democracy*, Environmental Challenges and Solutions 5, https://doi.org/10.1007/978-3-319-98307-3_4

so vast to us that we cannot imagine its finitude and fragility. As the physical and ecological limits of our home planet became more apparent in the twentieth century, Garrett Hardin argued that such limits necessitated a curtailment of fundamental human rights for the sake of protecting natural resources (Hardin 1968). In contrast, Elisabeth Mann Borgese made the case for a broad expansion of human rights, and a suprantional constitution to protect the global commons as the common heritage of the human race (Borgese 1965). This chapter will make the case that Hardin's reasoning was distorted by a false dichotomy between human rights and ecological sustainability, while the next chapter will make the case that Elisabeth Mann Borgese's vision transcends that false dichotomy.

The first realization of the earth's spherical shape is often attributed to the mathematicians of the Pythagorean school, though "the earliest writer whose work survives who. .. gives adequate proof of this fact, is Aristotle" (Evans 1998, p. 45). This discovery may have been the first intimation that it is not a world without limits, but rather a cyclical system in which, quite literally, what goes around comes around. A flat earth can be a limitless earth, but a spherical earth suggests a system in which changes affecting one part of the system will have reverberations throughout. Our recently acquired ability to see the earth from space should provide a cooling antidote to the political fever dream of perpetual conquest and the economic fever dream of perpetual growth. Nonetheless, we often tend to compartmentalize such visions of our spherical world and consider them only in relation to weather forecasts or the calculation of time zones. The rest of the time, we are likely to view the world around us ensconced in the same comfortable assumptions that guided the steps of our ancestors for thousands of years before anyone bothered seriously to contemplate and investigate the shape of the earth.

Occasionally a powerful metaphor can influence the mental habits of the human race. For this reason, it is worthwhile to take a look at the various metaphors that human civilizations have used to conceptualize our spherical earth and to envision its place in the cosmos. In the second century, the Egyptian astronomer Ptolemy built upon Greek science as he sought to identify the position of the spherical earth in relation to the sun, the moon, the planets, and the constellations. Ptolemy's model fixed the earth in the center of the universe and was structured on the supposition, which seemed to be confirmed by the empirical evidence available at the time, that all of the other heavenly bodies moved around the earth, while the earth itself was motionless. This vision of the earth and its place in the cosmos seemed to be confirmed by the daily experience of both astronomers and ordinary individuals. After the fourth century C.E., when Christianity was adopted as the official religion of the Roman Empire, the Ptolemaic paradigm offered the added benefit of seeming to confirm Christian doctrine. If man had been made in the image of the creator, and the destiny of man was imbued with a cosmic significance, then it only made sense that our world should rest at the very center of the universe. Although this model of our place in the cosmos is not literally true, it makes a metaphorical statement about the moral importance of human choices that is hard to ignore. The significance of this metaphor was appreciated by some modern authors, roughly three hundred years after the Ptolemaic worldview was laid to rest by the Copernican Revolution of the seventeenth century. In his 1938 play *Life of Galileo*, the Marxist playwright Bertolt Brecht shed light on the profound sense of disorientation that the Copernican Revolution created in the early seventeenth century. By the twentieth century, modern science had done far more than remove the human race from its imaginary perch in the center of the universe. It had also, in an age of total war, helped the military establishments of a growing number of nation states develop weapons of unprecedented destructive power. In a commencement address at Bennington University in 1970, the American satirist Kurt Vonnegut lamented the militarization of science in the twentieth century and its reduction of individual human beings to a state of contemptible insignificance. Contrasting this to an artistic view of our place in the universe, Vonnegut declared:

> The arts put man at the center of the universe, whether he belongs there or not. Military science, on the other hand, treats man as garbage — and his children, and his cities, too. Military science is probably right about the contemptibility of man in the vastness of the universe. Still — I deny that contemptibility, and I beg you to deny it, through the creation and appreciation of art. (Vonnegut 1974, p. 167)

Here Vonnegut urges us to embrace the spirit of the Ptolemaic worldview even if we cannot rationally accept it as a literal description of the material universe. His discovery of value in a scientifically obsolete cosmology is a rejection of what he calls "military science" but it does not call for a simplistic rejection of science itself. In an example of what the poet John Keats called "negative capability," Vonnegut is imploring us to embrace the moral framework implied by the Ptolemaic worldview even as he reminds us that a more dispiriting conception of our place in the universe is "probably right." Vonnegut's position is not a retreat into the wasteland of epistemic relativism, nor is it the promotion of fairy tales for the moral education of naïve children. It is nothing less than a sober recognition that the metaphors we use to describe our place in the world have a profound impact on our thinking and behavior. While we cannot choose to disregard the truth about the universe in which we live, we do have some leeway in our choice of those metaphors that will guide how we live.

In 1922, when the catastrophe of the Great War was still a fresh memory, Albert Einstein attempted to imagine how events on Earth would look to a thoughtful observer from the surface of the moon (Einstein 2007, pp. 121–122). Such an observer would see, Einstein contended, that the economic system of the world had grown through advances in communication and transportation, far beyond the boundaries of any single nation state. To prevent future conflicts, it would be necessary to create a political system that also transcended the boundaries of the nation state. Bringing his perspective back to the surface of the earth, Einstein argued that it was the duty of all people of goodwill and intelligence to advocate not merely for the wellbeing of their nation state but for the wellbeing of "the greater entity" on which all nation states depend (Einstein 2007, p. 122). Over the course of the twentieth century, more than one generation of scientists and intellectuals would attempt to create an evocative paradigm for that "entity" that Einstein viewed in his mind's eye from the surface of the moon. In the mid-twentieth century, the term "Spaceship

Earth" was promoted by the British economist Barbara Ward and later by the American inventor Buckminster Fuller. This conception fit nicely with the sense of technological wonder that dominated the 1950s and 1960s, but it faced a backlash in the 1970s. The Gaia hypothesis, promoted by James Lovelock, offered a more organic paradigm than "Spaceship Earth." It gained many adherents in the last three decades of the twentieth century, and alluded to the idea that the earth might be a self-regulating system similar to a living organism.

Both of these paradigms share the virtue of inviting us to think of the earth as a single system, but each is limited by some irreducible drawbacks. As "Spaceship Earth" emerged during the intense technological competition of the Cold War, it is perhaps not surprising that this metaphor suggests a starkly technocratic approach to our ecological situation (Deese 2009). Perhaps as a reaction to the technocratic connotations of the Spaceship Earth metaphor, many advocates of global environmentalism embraced the Gaia hypothesis in the early 1970s, making it "the most widely discussed scientific metaphor of the Age of Ecology" (Poole 2008, p. 170). Lovelock has been adamant that the Gaia hypothesis was not intended to be a religious idea (Lovelock 1995, p. 192). However, both supporters and critics have tended to dwell on its religious implications, beginning with the primeval Greek goddess who furnishes its name. This association of the Gaia hypothesis with religion, whether warranted or not, has alienated both rational skeptics who are repelled by any whiff of religious dogma, and votaries of traditional religions who may see it as a competing faith. On the secular end of this spectrum, the biologist and noted atheist Richard Dawkins has ridiculed the Gaia hypothesis for its quasi-religious assumptions about a sense of purpose in nature (Ruse 2013, p. 27, 149). On the religious end of this spectrum, the pagan roots of this metaphor have bolstered arguments, promulgated by evangelical leaders such as "Pat Robertson, Jerry Falwell, and Tim LaHaye. .. [who] cast environmentalism as a New Age religion" incompatible with the values of Abrahamic monotheism (Howe 2016, p. 126).

In the twentieth century, another metaphor emerged to describe the earth that is remarkable in both its simplicity and its accuracy. This is the metaphor of the earth as an oasis. The word "oasis" actually entered the English language from demotic Egyptian as a term to describe a small space with water and life in a vast expanse of desert. Aldous Huxley put the term to good use when he revised the most famous dictum from Voltaire's satirical novel *Candide*. Where Voltaire ended his novel with the protagonist's declaration that "Il faut cultiver notre jardin" ("We must cultivate our garden"), Huxley was fond of saying "Il faut cultiver notre oasis" (Deese 2015, p. 175). For Huxley, this vision of care for our ecological inheritance began as an attachment to the local landscapes that he loved, especially the deserts of the American southwest. By the late 1940s, however, Aldous Huxley increasingly emphasized that the oasis of the entire Earth required care and protection. In a letter to his son Matthew in 1948, Aldous declared that, "we must all get together on a world conservation policy upon which the various nations can possibly agree, because having enough to eat is the one thing that concerns everybody equally" (Deese 2015, p. 154–155). After seeing the earth rise above the lunar horizon in December of 1968, the Apollo astronaut Jim Lovell described our home planet as "a grand oasis" (McKibben 2010, p. 2). The pioneering ocean conservationist Jacques Cousteau also saw the value of the metaphor of the oasis, and employed it in his 1972 television program "Oasis in Space." Seventeen years later, the award-winning earth scientist Preston Cloud embraced this term as well and used it as the title of his final book (Cloud 1989).

In the early twenty-first century, when the American astronaut Ron Garan gazed down at the earth from the International Space Station, he also saw the earth as an oasis in need of care and protection:

> . . . [A]s I looked down at the Earth—this stunning, fragile oasis, this island that has been given to us, and has protected all life from the harshness of space—a sadness came over me, and I was hit in the gut with an undeniable, sobering contradiction. In spite of the overwhelming beauty of this scene, serious inequity exists on the apparent paradise we have been given. I couldn't help thinking of the nearly one billion people who don't have clean water to drink, the countless number who go to bed hungry every night, the social injustice, conflicts, and poverty that remain pervasive across the planet. (Ron Garan 2015. *The Orbital Perspective*, p. 4.)

Garan's description of the Earth as a "stunning, fragile oasis" is both poetic and scientifically sound. A survey of the other planets in our solar system underlines the power and accuracy of the oasis metaphor. Our closest planetary neighbors, Venus and Mars, illustrate how relatively moderate variations in size, atmosphere, and proximity to the sun can produce much greater variability in the conditions for life. In terms of its size, position, and atmospheric composition, Earth is in what astronomers have called the "Goldilocks Zone" because its conditions for life are, as Goldilocks would say, "just right." In practical terms, this means we are indeed living on an oasis in space, and we must guard it from the threats of waste, pollution, and war. This realization that the planet on which we live is a miraculous gift has been affirmed by modern science, but it was intuited by many religious traditions for millennia. The concept of *wouncage* common to the traditions of several Native American tribes of the Great Plains, is one example of this ancient intuition. As the journalist Paul VanDevelder writes, "Translated from the Crow, Mandan, or Sioux, *wouncage* is today known as Sacred Trust." It expresses "the reverence of the people of a tribe or community for the originating force that moves the wind, that brings the clouds, that carries the rain, that falls to the grass, that feeds the buffalo to nourish the man" (Vandevelder 2009, p. 244). As with Ron Garan's revelation of our "fragile oasis" from the vantage point of the International Space Station, the concept of *wouncage* does not draw a line between nature and humanity to privilege one over the other. Rather, it recognizes that human wellbeing and respect for the grandeur and integrity of nonhuman nature are inextricably linked.

Evolutionary biology indicates that we are part of nature, and that we emerged, over hundreds of millions of years, from what Darwin described as a "tangled bank" of ecological relationships. However, evolutionary biology runs counter to an older set of religious traditions which teach that we are the special creation made in the image of the supreme deity and thus entirely distinct from the rest of life on earth. Although we may think of science as distinct from religion, this religious legacy shaped the thinking of Descartes and Bacon and generations of scientists who followed their lead. For Descartes, nonhuman animals were nothing more than automatons, devoid of any consciousness and ready to be exploited for whatever purpose human beings could imagine (Hill 2011. p. 112). For Bacon, the goal of science was to reestablish, through the steady advance of technology, the complete dominion that Adam had enjoyed over nature before the Fall (Almond 1999. p. 35). As founders of the broad traditions of Cartesian rationalism and Baconian empiricism, these thinkers wove a binary opposition between humanity and nature into Western conceptions of science.

When the environmentalist movement emerged in the twentieth century, it rejected the notion of human dominion over nature, but the binary opposition between humanity and nature remained nonetheless. The persistence of this binary opposition created false dichotomies in the analysis of many leading environmentalists concerning both politics and technology. In the realm of technology, many environmentalists exalted the concept of the pristine wilderness, and condemned the advances in technology as a departure from the innocence of Eden. These false dichotomies had their greatest influence on environmental discourse in the 1970s and 1980s, but they have colored public perception of the environmentalist movement ever since.

In the realm of politics, human rights and environmental sustainability were placed in stark opposition to one another, especially during the first three decades after 1945, when Malthusian fears about overpopulation, especially in developing countries, had heightened influence on the thinking of academics and policy experts in the United States (Robertson 2012). The growing dichotomy between environmental sustainability and basic human rights was exemplified by the enormous influence of Garrett Hardin's 1968 essay "The Tragedy of the Commons." This seminal essay advanced the central argument that state power should be employed to "close the commons of human reproduction." While birth control advocates such as Annie Besant and later Margaret Sanger had advocated for more than a century that birth control and information about how to use it should be freely available to women, Hardin's essay advocated an authoritarian program in which the government would make and enforce decisions about family size. Hardin's decision to privilege state power over human rights was justified as a necessity for human survival, but demographic trends since the publication of "The Tragedy of the Commons" indicate that his draconian proposals were both impractical and misguided. The idea that freedom and the innovation that it makes possible would disprove Malthus was intuited by Emery Reves in *A Democratic Manifesto*, and it has been gathering force in the decades following the population scare of the 1970s (1942. p. 39).

Where Hardin posited less individual freedom as the answer to the Malthusian crisis that concerned so many in the late 1960s, it turns out that greater freedom, especially the freedom for women to acquire a secondary or tertiary education and to enter the workforce, has been the greatest check to runaway population growth (Sanderson 2004, p. 240). Instead of abrogating the fundamental rights of individuals, as Hardin proposed, environmentalists should advocate full legal equality for women across the world, and achieve the goal that Elizabeth Cady Stanton articulated in 1848 when she declared at Seneca Falls that "all men and women are created equal" and that no woman should be denied access to "her inalienable right to the elective franchise" nor to "the facilities for obtaining a thorough education" (Stanton 2015. pp. 6, 10).

In addition to advocating the imposition of statist policies to control family size, Hardin advocated eugenics and proposed severe restrictions on immigration that would be based on race and ethnicity. In the 1970s, Hardin advocated the forced sterilization of welfare recipients as part of the "coercive phase" of population control which he expected would soon become necessary (Hardin 1973, p. 199). In 1997, he declared in an interview in the anti-immigration journal *The Social Contract* that "My position is that this idea of a multiethnic society is a disaster. That's what we've got in Central Europe, and in Central Africa. A multiethnic society is insanity. I think we should restrict immigration for that reason" (*The Social Contract*. Fall 1997). Casting himself as a brave iconoclast, Hardin took aim at two propositions that he viewed as absurd: "(1) there will come a day when all humanity is united into 'One World' bound by a single political plan and (2) that human ethical principles prohibit discrimination among hereditary types of people." Hardin argued that the acceptance of such positions could possibly lead not only to the destruction of the United States through increased "Latino immigration" but even to the dissolution of civilization itself. Hardin warned his readers that the "suicidal" pro-immigration policies recommended by the Ford Foundation, might "succeed in creating a chaotic NorteAmericano Central" and closed his essay with this dire speculation: "The human species may not self-destruct; but what we like to call 'human civilization' may. This is a sobering thought for scholars to ponder" (Hardin 1999). While his essay "The Tragedy of the Commons" has earned him a lasting place in the history of environmental thought, Hardin's long record of statements advocating eugenics and targeting certain racial and ethnic groups as undesirable has been less remarked upon. However, the Southern Poverty Law Center, a civil rights organization that carefully tracks racist groups in the United States, has categorized Hardin as a White Nationalist (SPLC 2017).

Although Hardin's intellectual credentials would seem to put some distance between his work and the rantings of Klansmen or neo-Nazis, his endorsement of eugenics and his conviction that "a multi-ethnic society is insanity" point to troubling contradictions that have haunted the environmentalist movement for a very long time. Opponents of environmentalism frequently point to the advocacy of eugenics by figures such as Julian Huxley as part of a rhetorical strategy to associate contemporary environmentalism with Nazism, in spite of the fact that Julian Huxley was one of the earliest and most vocal critics of Nazi racial policies in the 1930s (Deese 2015, p. 74). Although the most partisan attacks on the history of environmentalism have been tendentious and misleading, it is still necessary for contemporary environmentalists to acknowledge the strains of racist and antidemocratic thought that have tarnished the movement over the past century. As the writer and legal scholar Jedediah Purdy has observed, "Some of the awkwardness of environmental politics" is a consequence of the fact "that it lays claim to worldwide problems, but brings to them some of the cultural habits of a much more parochial, and sometimes nastier, movement" (Purdy 2015).

Garret Hardin's seminal essay "The Tragedy of the Commons" was published in December of 1968, the same month that over a billion people around the world saw the first televised image of the earth from the vantage point of the Apollo 8 mission (Poole 2008, p. 26). This essay has had such an enormous impact on environmental discourse because its central metaphor of shepherds depleting a common pasture expresses a disturbing truth: where nature had once surrounded us, we have come to surround nature. For most of human history, our tribes and settlements, even our cities and empires, have amounted to little more than some colonies of ticks on the back of a huge living entity that we have at one time or another called nature, mother earth, Gaia, etc. By the second half of the twentieth century, however, it was clear to any informed observer that the human race had become much more than a few colonies of ticks. Our networks of trade and communication encircled the entire earth, our appetite for the earth's resources and our ability to extract them continued to increase exponentially, and we even possessed, and continued to refine, the ability to annihilate life on earth with nuclear weapons.

"The Tragedy of the Commons" begins, in fact, with a discussion of the nuclear arms race. Echoing an argument that had been made by a number of Manhattan Project scientists immediately after World War Two, Hardin points to the findings of Jerome Wiesner and Herbert York that the U.S. and Soviet blocs in the nuclear arms race were now "confronted by the dilemma of steadily increasing military power and steadily decreasing national security. *It is our considered professional judgment that this dilemma has no technical solution*. If the great powers continue to look for solutions in the area of science and technology only, the result will be to worsen the situation." The original purpose of Wiesner and Herbert's article had been to argue that anti-ballistic missiles would not offer a viable solution to the nuclear arms race. Although he does not take a position of his own on nuclear weapons or anti-ballistic missile research, Hardin finds their observation particularly astute because of its implicit message that some problems do not afford a technical solution. He goes on to argue, that the stresses which a growing human population is placing on the earth's ecological carrying capacity is precisely this kind of problem. He reasons that we will not miraculously escape it by developing new crop strains or new sources of energy, but will instead be required to embrace fundamentally different modes of thinking and behavior.

By starting his message with a discussion of the nuclear arms race, however, Hardin alludes to the volatile relationship between human civilization and nature in the age of nuclear technology. We have nature surrounded, not so much like a group of shepherds surrounding a common field, but more like an army of heavily armed police surrounding a hostage situation. At any moment, things could go terribly wrong, and the world could become a radioactive graveyard. By using the analogy of the nuclear arms race to establish his premise that some problems do not have technical solutions, Hardin signals that his use of the word "tragedy" here is anything but casual. As he sees it, the ecological crises engendered by overpopulation are every bit as serious as the political crisis of the nuclear arms. In fact, the destabilizing effects of overpopulation made the danger of nuclear war that much greater in Hardin's view. This premise was shared by other Neo-Malthusian thinkers in the late 1960s, and perhaps found its most compact expression in the title of Paul Ehrlich's 1968 bestseller *The Population Bomb*.

Although Hardin's essay was primarily concerned with human population growth, the analogy of a common pasture that Hardin used to frame his argument was a perfect match for the historical moment. Unlike the schoolroom globe with its clear delineation of national boundaries, the earth seen from space was a startling spectacle of beauty, fragility, and singularity. As it appeared above the horizon of the lifeless moon on Christmas Eve 1968, the most striking and dynamic features of our planet were in fact its oceans and its atmosphere, or what has come to be called the global commons. In this historical moment, when the photographic evidence that our world is a single unit became as ubiquitous as the face of a celebrity, Hardin's essay gave scientists, economists, and public intellectuals a very handy tool to talk about that unity (Deese 2008).

Hardin borrowed the idea of the commons, or a common plot of pastureland owned by no one in particular and shared by shepherds who live on its periphery, from British and economic and social history. The closing of the commons through the privatization of plots by large landholders forced peasant farmers to seek work in towns and cities, was a process that took place over several centuries. As early as the sixteenth century, Sir Thomas More had lamented the closing of commons in the English countryside and the poverty and desperation that it engendered, but the practice took on a new momentum during the Industrial Revolution. Large landowners, eager to maximize the value of their holdings, closed common areas and displaced the peasantry who had depended on the unclaimed land for grazing livestock and small-scale farming. These displaced peasants, as they poured into population centers in the English midlands such as Manchester became the workforce for a newly centralized system of factory production. As that system grew more rapidly through the displacement of older water-powered mills by new machinery that was powered by coal-fired steam engines, the closing of the commons became imperative to sustaining the momentum of Britain's industrial revolution. By the early nineteenth century, the closing of the commons went from being a necessary evil to a positive good in the eyes of some economists.

In 1833, the Oxford economist William Forster Lloyd coined the metaphor that Hardin would later make famous. Lloyd compared the ill-fed cattle that subsisted on common lands to the more robust specimens found on the large estates of wealthy landowners: "Why are the cattle on the common so puny and stunted? Why is the common itself so bare worn, and cropped so differently from the enjoining inclosures?" (Hardin 1968, p. 31). Although by this point those lands that remained common were some of the poorest in England, Lloyd explained the difference in the health of the cattle entirely in terms of the owners' motivation in each case. The landless farmers who grazed their cattle on the common lands took no heed of the additional burden that they placed on the local vegetation with each new head of livestock, but the landowning farmers had to keep a careful account of how much cost each new head of cattle would impose upon their pastures. Thus, the common lands were overgrazed the livestock upon them were "puny and stunted," while the privately-owned lands in the "inclosures" were thick with vegetation and populated by cattle that were conspicuously healthier and better fed. Where three centuries earlier Thomas More had blamed the process of enclosure for the poverty of peasant farmers, William Forster Lloyd turned that argument on its head. It was the continued existence of the commons itself which exacerbated the poverty of the peasants by encouraging the over-breeding of cattle and depleting the value of the land. If all land were privatized, the misuse of land seen on the commons would be eliminated because landowners would rationally see it in their self-interest to avoid the overgrazing of their *own* lands. As long as the commons existed, Lloyd reasoned, peasants who grazed livestock there would be motivated to overgraze it with their own cattle because they knew that if they did not, somebody else would and the commons would be overgrazed in the end, regardless of their choice. In such a case, the rational self-interest of every cattle-driving peasant on the commons led inexorably to the destruction of the commons itself, even if they could see it coming.

In 1968, Hardin took this argument about the pasturelands of the British countryside and made it global. Where William Forster Lloyd had concentrated on the impact of cattle populations on grasslands, Hardin considered an issue that was both more expansive and more personal: the impact on the earth's resources by human population growth. Abstracting the central principle from Lloyd's 1833 treatise, Hardin reasoned that, "Ruin is the destination toward which all men rush, each pursuing his own best interest in a society that believes in the freedom of the commons. Freedom in a commons brings ruin to all" (1968, p. 20). In this essay, the freedom that concerned Hardin the most was the freedom of human beings to breed, not cattle, but their own progeny. While W.F. Lloyd had recommended the complete closure of common pastureland in the early nineteenth century, Hardin called for restriction on a much more fundamental human freedom in the late twentieth century. Declaring that "freedom to breed is intolerable," Hardin reasoned that a recent United Nations resolution on human rights and family planning was fatally flawed for including that freedom: "To couple the concept of freedom to breed with the belief that everyone born has an equal right to the commons is to lock the world into a tragic course of action" (1968, p. 24).

Hardin characterized his criticism of the United Nations and its position on reproductive rights as an act of remarkable moral courage in defiance of the liberal consensus of the era, and compared himself to a citizen in seventeenth-century Massachusetts who might dare to deny the existence of witches (1968, p. 24). In fact, his stark dichotomization of human rights and sustainable population growth had roots in the conservative backlash against some of the egalitarian ideals that had emerged from the Enlightenment, and the Declaration of the Rights of Man and the Citizen that marked the first phase of the French Revolution. Hardin's oppositional dichotomy between a society which extends basic dignity and human rights to its population and a society that can feed itself echoes the stark dichotomy that Thomas Malthus outlined in his *An Essay on the Principle of Population* in 1798. When Malthus calculated that human population growth, left unchecked by war, famine, and pestilence, would soon outstrip available food supplies, his initial aim had been to critique the utopian philosophy of the mathematician and philosopher Nicolas de Caritat, marquis de Condorcet.

An advocate of equal rights for women and for all races, Condorcet had argued in his *Sketch for a Historical Picture of the Progress of the Human Mind*, that the possibilities for human progress through universal education, democracy, and the application of scientific knowledge were virtually unlimited. Remarkably, he wrote this optimistic treatise in prison, after he had been arrested and condemned to death by Robespierre. Though Condorcet died in prison soon after completing it, the ideas in his last treatise proved influential. As historian Paul R. Josephson has put it, Condorcet's ideas reflected the revolutionary vision of science and human potential that "gained currency in Europe and North America during the Enlightenment" and his final treatise "best expressed this relation of progress, society, and the natural world" (2005, p. 10). Condorcet's treatise soon found its way to England where its ideas were refined and promoted by the English journalist and political philosopher William Godwin.

Since it was composed after a decade in which France had been wracked by famine, we should not be surprised that a major tenet of Condercet's treatise was his claim that science and engineering could reduce hunger by steadily improving agricultural methods. The condemned Marquis predicted that in the future, "not only the same amount of land be able to feed more people; but each of them, with less labor, will be employed more productively and will be able to satisfy his needs better" (Josephson 2005, p. 11). This message of plenty had wide appeal, but Condorcet's connection, even to the moderate Girondin faction of the French Revolution, was enough to taint the provenance of his ideas for many in England. Among Godwin's compatriots, there were many who saw the political implications of the French Revolution as an unmitigated catastrophe. *An Essay on Population* by the Anglican cleric Thomas Malthus was part of this tide of reaction. Malthus took aim at the notion that the restructuring of society could eliminate poverty. Malthus penned his several versions of this essay between 1798 and 1826 as he attempted to debunk the utopian reformism of both Condorcet and Godwin. The core of his argument was brutal in its simplicity. Because population increases exponentially while food supply can only increase arithmetically, the number of mouths to feed will necessarily outstrip the supply of food with which to feed them. Therefore, famine, pestilence, and war are actually necessary checks on the excess population. If the schemes of reformers such as Condorcet and Godwin for improving society could actually succeed in eliminating poverty, war, and disease, any nation that adopted them would inevitably be brought down by overpopulation and famine.

The predictions of Thomas Malthus, based as they are on his premise that the growth of food supply is arithmetic while human population growth is "geometric," are usually treated as a scientific hypothesis. This is the best way to evaluate his specific claims, but it is not the best way to understand the origins and impact of Malthusianism, or "neo-Malthusianism" in the history of Western discourse on population and poverty since the eighteenth century. Within the framework of that discourse, the circumstances under which Thomas Malthus wrote and published "An Essay on the Principle Population" are very revealing. By 1798, when Malthus wrote his first edition of the essay, the French Revolution had elicited intense admiration from an entire generation of British artists, writers, and intellectuals, even as it inspired horror among conservatives such as the Irish Protestant Member of Parliament Edmund Burke. The Reign of Terror that reached its nadir in the early 1790s had caused many to despair that the goals and methods of the revolutionaries had gone badly off course, but Burke argued that the French Revolution had been doomed from the outset because it had been based on the premise that scientists and intellectuals could displace the established authority of the clergy and aristocracy in France. Commenting on events in Paris, Burke lamented the passing of aristocratic and royal authority and saw little virtue in the rise of an intellectual and managerial class in European society. He grimly declared, "the age of chivalry is gone. That of sophisters, economists, and calculators has succeeded; and the glory of Europe is extinguished for ever" (Burke [1790] (2012), pp. 74–75). As an Anglican minister, Malthus shared Burke's revulsion to the idea of progress through science and reason that French intellectuals had advanced during the first phase of the French Revolution. To be more precise, his rejection of the French Revolution was not directed at the disciples of Rousseau who promulgated the Terror, but rather at liberal progressives such as Condorcet who actually became victims of that period of paranoia and bloodletting. By the late 1790s, the optimistic vision of Condorcet had inspired English intellectuals such as William Godwin to argue that a more rational and democratic structure for society could eliminate poverty and move the human race toward the creation of a perfect, or nearly perfect, society. Of course, many would criticize such thinking as unrealistic, but for religious conservatives such as Burke and Malthus, it was something much worse: a blasphemy against the doctrine of original sin. Consequently, the Malthusian attack on the utopian idealism of Condorcet and Godwin stemmed from a deep well of religious feeling, even if it presented itself in the language of a scientific hypothesis. The unscientific nature of Malthusian thought is implicit in the framing of his thought experiment. By the time Malthus set his pen to paper, the condom had become commonplace throughout Europe; nonetheless, the potential benefits of birth control are never considered in "An Essay on the Principle of Population" (Collier 2007, p. 124) Just as medieval parsons had accepted poverty as a divine necessity and argued that the best hope for the poor remained in the next world, Malthus seemed to affirm implicitly, through his relegation of birth control to the realm of "vice," that poverty and destitution must be a permanent aspect of human existence.

The antipathy to Condorcet that animated Malthus is also present in the work of Garrett Hardin. Like Malthus, Hardin regarded Condorcet as hopelessly naïve in his visions of progress and universal human rights. Hardin describes the ill-fated reformer as an "ugly duckling of the aristocracy," who dabbled with no more than a passable competence in both mathematics and political theory. In the case of William Godwin, Hardin's judgment is even more dismissive. After reviewing the key points of Godwin's life, including his unconventional life with Mary Wollstonecraft and their daughter's ill-starred marriage to Percy Bysshe Shelley, Hardin composes the following epitaph for Godwin: "Father of the Author of Frankenstein, and Irritant that Produced Malthus" (1993, p. 21).

In Hardin's estimation, "Condorcet and his followers have had more influence on the climate of opinion in our time than has Malthus," for the simple reason that, "Optimism is more attractive than pessimism" (1993, p. 24). Hardin argued that the progressive spirit of these two Enlightenment thinkers deserved to be displaced by a more clear-eyed and frankly Malthusian view of human affairs. Such visions of human progress may have had the benefit of spurring innovation and economic growth since the beginning of the nineteenth century, but by the closing decades of the twentieth century, Hardin suggested that the world could more clearly see the limits that Malthus had envisioned:

> Our increasing anxiety about the depletion (of material wealth) and the increase of pollution (by material wastes) makes us wonder whether we are not at last approaching "the limits of perfectibility" of our materialistic world. Though not decisive, the present trend is clear enough to make some of us have second thoughts about our much vaunted "progress" (1993, p. 25).

The world's population had just reached one billion when Malthus and Godwin were publishing their essays, and Godwin's optimism more closely reflected the zeitgeist of the nineteenth century. By 1968, when Hardin published "The Tragedy of the Commons," it had more than tripled and was on track to double again before the end of the twentieth century. Given this dynamic growth in human population, coupled with the mounting evidence of the finite nature of the global commons, made vivid by the NASA broadcast of the "earthrise" images in December of that year, it is not surprising that the zeitgeist of the late 1960s and early 1970s turned, just as Hardin had predicted, away from the liberal optimism of Condorcet and Godwin, and toward the dour life-raft ethos of Thomas Malthus. In the twenty-first century, some political scientists have critiqued the antidemocratic bent of Hardin and his contemporaries. In his study *Democracy and Global Warming*, Barry Holden describes the worldview that Hardin articulated as the "survivalist stage in green political thought" and observes that it was "characterized by strongly anti-democratic" tendencies (2002, p. 6–7).

In the decades since Hardin first published "The Tragedy of the Commons," the concepts that he articulated in that essay have been explored and altered by more than one generation of scholars and researchers, yielding such concepts as the "Comedy of the Commons" coined by and Carol Rose and the "Tragedy of the Anticommons" described by Michal Heller (Deese 2008). The most thorough research on the issues Hardin raised was conducted by the Nobel Prize winning economist Elinor Ostrom. Her research led her to conclude that Hardin's assumptions were only confirmed in situations where participants acted anonymously and were neither visible nor accountable to each other: "The predictions of noncooperative game theory are roughly supported only when participants in a laboratory experiment do not know the reputation of the others involved in a common-pool resource dilemma and cannot communicate with them" (Ostrom 2009). In more transparent situations, Ostrom concluded that the creation of agreed upon norms and sanctions had the best track record for the management and preservation of common-pool resources:

> . . . [W]hen subjects communicate face-to-face, they frequently agree on joint strategies and keep to their agreements – substantially increasing their net returns. Further, communication to decide on and design a sanctioning system enables those choosing this option to achieve close to optimal returns (Ostrom 2009).

Ostrom's conclusions in the field of economics have found support in the illuminating research of Barry Holden and Judith Shapiro. Working respectively in the fields of political science and history, Holden and Shapiro have demonstrated with great clarity that freedom of speech and respect for human rights are essential to transparency and accountability. In turn, transparency and accountability are, as Elinor Ostrom has shown, essential to the sustainable management of common-pool resources. The history of common poor resources (CPRs) does not point to the "inexorable" process of ruin that Hardin predicted. As Elinor Ostrom and Robert Keohane have observed, a wide range of "successfully governed CPRs have survived for centuries relying on self-monitoring and self-enforcing patterns of human interaction" (1995, p. 1). Paul R. Josephson, who has surveyed the environmental impacts of various types of contemporary political regimes, advances the conclusion that democratic or pluralist states have a superior environmental record to authoritarian regimes, "largely because of broader access by citizens to information about the appropriate path of development, and because of the creation of legal, scientific, and other institutions to mitigate environmental problems" (2005, p. 20). The work of these scholars provides ample evidence that Garrett Hardin's tendency to embrace authoritarian solutions to environmental problems was predicated on a false dichotomy between human rights and environmental sustainability. In fact, the regimes that have the greatest respect for human rights are also likely to have the greatest respect for the environment. Conversely, regimes that deny the basic human rights and electoral accountability to their citizenry tend to abuse the environment as much as they abuse their own people. Power, when it flows one way, carries poison with it.

Another false dichotomy that has often distorted environmental discourse is the stark opposition between technology and nature. This is a split that has deep roots in Western religion and philosophy and which is still apparent in our collective fantasies about the future. Since at least the seventeenth century, when Francis Bacon first sketched his vision of a technocratic utopia in the *New Atlantis*, there have been those who have dreamed of using science and technology as tools for attaining the complete control of nature. In *Silent Spring* Rachel Carson assaulted such dreams with withering eloquence when she declared: "The 'control of nature' is a phrase conceived in arrogance, born of the Neanderthal age of biology and philosophy, when it was supposed that nature exists for the convenience of man." And yet even Francis Bacon, who dreamed that human mastery over nature would "restore the estate of Adam" and return us to the Garden of Eden, seemed to intuit that that the complete control of nature was a logical impossibility. Bacon presents a paradoxical aphorism in his 1620 treatise on the scientific method, *Novum Organum*: "Nature, to be commanded, must be obeyed." What we might call "Bacon's Paradox" points to an enduring irony that will always haunt any human attempt to control nature.

Bacon's Paradox is irreducible because our species, being dependent on the use of tools, we cannot escape *into* nature, and, being the animals ourselves, we cannot escape *from* nature. These alternative fantasies of escaping into nature and escaping from nature have a long history for the human race, but the longing for both of these impossible outcomes has become particularly acute in the industrial age. Our popular culture reflects the intense power of these fantasies. The dream of escaping *into* nature, which has fueled romanticism for over two centuries, was remarkably well expressed in what became the highest grossing film in history, James Cameron's 2009 blockbuster *Avatar*. The dream of escaping *from* nature has also found expression in popular culture, as exemplified by the premise of Christopher Nolan's critically acclaimed 2014 science fiction film, *Interstellar*.

In *Avatar*, the devastating impact of imperialism and extractive industries on indigenous people is depicted in a science fiction scenario set in the twenty-second century. In this case, the imperialists are earthlings of a distinctly American cast who have come to extract a miraculous mineral called "unobtainium" from a verdant moon called Pandora in the Alpha Centauri system. Complicating this extraction is the existence of a highly intelligent race of seven-foot-tall blue humanoids on Pandora known as the Na'vi. The protagonist of *Avatar*, a young paraplegic Marine by the name of Jake Sully, is recruited by a group of military and corporate operatives to infiltrate the Na'vi and help destroy their society from the inside out, so that mining operations for unobtanium can be accelerated without further complications.

The plot structure of *Avatar* resembles Aldous Huxley's *Island*. Like Jake Sully in *Avatar*, the character of Will Farnaby in *Island* is enchanted and inspired by an ecologically sustainable society at the same time that he is covertly aiding those who want to destroy that society for the sake of plundering its resources. However, *Island* presents an ecologically sustainable society that has consciously adopted certain elements of industrial technology while rejecting others. The people on Huxley's utopian isle have had a deliberate impact on the landscape around them, and their technology is sophisticated although carefully applied. Describing the artfully terraced hillsides of Pala, Huxley declares, "Nature here was no longer merely natural.. ." (Huxley [1962] (2002), p. 23). *Avatar* presents a very different scenario in this respect. The Na'vi on the verdant moon Pandora have only the simplest tools and, rather than reshaping the landscape as farmers do, they are part of the landscape. In their untutored simplicity and oneness with the landscape, the Na'vi owe less to Aldous Huxley's vision of pacifism and ecologically sustainable technology, and more to Rousseau's idealized vision of the noble savage.

Echoing a theme as old as empire itself, James Cameron's *Avatar* presents a situation where a young white male "goes native" – first by falling in love with a beautiful native woman named Neytiri, and then by learning to embrace the ecologically harmonious culture of the native people. Although Pandora itself resembles a combination of a tropical rainforest and a painting by the fantasy artist Roger Dean (famous for his album covers for the rock group Yes), the planet's inhabitants talk, dress, and fight with mannerisms that closely resemble the noble Native Americans depicted in the American Western films such as *Dances with Wolves* and *A Man called Horse*. These blue indigenous people ride with the courage and skill of Lakota warriors, although instead of riding horses they mount the backs of huge flying reptiles. *Avatar* is not content to borrow the old cliché of the Native American warrior who becomes one with his horse, however. When these riders take to the skies, each literally connects his own brain to the brain of his flying lizard, through a cord of nerves that extends from the end of his braided hair. This literal connection is not limited to riders and reptiles: on this planet, lovers can also join their nervous systems to one another in this way, and all of the Na'vi can connect their nervous systems to the vast ecosystem of Pandora by connecting their pony tails to a giant arboreal nerve network called the Tree of Souls.

When Jake Sully learns that his corporate and military paymasters are preparing a direct strike upon the Tree of Souls, he shifts his allegiance completely to the side of the Na'vi and uses his considerable skills as soldier and battlefield orator to unite every Na'vi tribe on Pandora against the military forces that he had previously served. These battle scenes, which depict Na'vi counter attacks against futuristic gunships in a tropical ravine, recall the imagery of the post-Vietnam antiwar films of Oliver Stone and Francis Ford Coppola, complete with a demented southern martinet, Colonel Miles Quaritch, directing the helicopter assaults. After he leads the Na'vi to their victory over Quaritch and his forces, Jake Sully cuts all ties to his former masters and to the human race itself, making a complete and irrevocable transformation into a Na'vi warrior.

Jake's path from U.S. Marine grunt to blue-skinned warrior on the moon of Pandora is a miraculous transformation, and it forms the spine of an incredibly entertaining film. In *Avatar*, writer and director James Cameron created a myth that speaks to a profound set of desires shared by the citizens not only of the United States, but of industrial societies around the world. The global success of *Avatar* indicates that the concept of the "noble savage" first articulated by Montaigne in the seventeenth century has pervasive appeal among the rapidly growing population of people around the world who possess the leisure time and disposable income to visit a shopping mall or Imax venue and watch a three hour 3-D film about blue warriors taking on the military industrial complex. Of course, *Avatar*, with its fast-food toy tie-ins, theme park spin-offs, and inevitable sequels, is a product of the very culture of ubiquitous consumerism that its protagonist, Jake Sully, explicitly rejects. When Jake reaches the conclusion that the earthlings have virtually nothing of value to offer the inhabitants of Pandora, he points directly to the emptiness of consumer culture. The Na'vi, he concludes are, "not gonna give up their home. They're not gonna make a deal. For—for what? Light beer and blue jeans? There's nothing that we have that they want" (*Avatar* 2009). The fact that a blockbuster film such as *Avatar* is a product of the same cultural and economic forces that have given us light beer and blue jeans does not render the film meaningless or hypocritical, but it does point to a knot of contradictions in our thinking about technology, nature, and culture that cannot be easily untangled. The eco-friendly subtext of *Avatar* posits that the Na'vi are a perfect society, because they live in perfect harmony with their environment. Because they are the products of computer generated imagery who inhabit an imaginary world light-years from ours, the viewer has no compelling reason, on the face of it, to question this premise. On earth however, such perfect harmony with the environment has never been achieved. Even hunter-gatherer societies have driven species into extinction through excessive hunting, and no society of human beings has lived in a relationship with the natural world that did not include threats to its own wellbeing from such natural calamities as droughts, powerful storms, and disease – none of which appear to be a problem on the idyllic world of Pandora. Unlike the Na'vi, human beings have spent most of their history as species regarding the forces of nature as something that could kill them. It requires an industrial age of vaccinations, antibiotics and climate-controlled shopping malls to generate an audience for a fantasy in which the hero rejects all the trappings of industrial society for the life of a nearly naked tribesman in a tropical rainforest. It's appealing for comfortable movie-goers to imagine roughing it as Jake Sully does in the Na'vi rainforest, especially when life in that luminous blue rainforest is entirely free of mosquitoes, snake bites, and tropical diseases.

As a high-tech consumerist *cri de coeur* against high-tech consumerism, *Avatar* is hardly unique. However, the same contradictions that surround its success as a blockbuster movie also infect its plot. Jake Sully's rejection of human interference with nature through advanced technology would never be possible without – and this should surprise no one – a great deal of human interference with nature through advanced technology. Jake can only become a Na'vi through the creation of a genetically engineered Na'vi avatar which his brain controls through a complex bio-electronic interface. This premise points to further borrowing by James Cameron from earlier works of science fiction and fantasy. Just as the fantastic landscape of Pandora owes a great deal to album covers of the 1970s, the most central idea of *Avatar*, the idea of a cable connecting one's own brain to other brains and to a global network, belongs to the cyberpunk genre of the 1980s. Where the novelist William Gibson envisioned a highly technological world in which his characters could "jack in" and directly connect their brains to a global network of computers, the Na'vi can naturally connect themselves to a global network of living things via the Tree of Souls. Ultimately, our young Marine will join them in this apotheosis of harmony with nature, but only because the military-industrial complex that he has hitherto served has created for him a marvelously seamless Frankenstein body and brain that perfectly mimics the natural body and brain of a Na'vi warrior. In joining the Na'vi, he casts aside his natural body to spend the rest of his life in a completely synthetic simulation that has been grown in a laboratory.

Nearly four centuries after Francis Bacon opined that technology would one day allow mankind to restore the estate of Adam, James Cameron depicted that achievement in Avatar: through the technological miracles of space travel, synthetic biology, and bio-electronics, the wounded and spiritually lost Jake Sully is able to infiltrate an extraterrestrial Eden, find his Eve, and then free himself from the sense of alienation from nature that has haunted every child of Adam since the Fall. The romantic sentiments that permeate Cameron's vision are derived from the Romantic tradition that began in the late eighteenth century, but the audacious vision of transformative technology that drives Cameron's narrative forward bear the distinct genetic stamp of Bacon's vision as expressed in such works as *Novum Organum* and

*The New Atlantis*. In creating a science fiction idyll with such romantic themes, Cameron has not refuted Bacon's vision of the relationship between technology and nature but has instead provided a corollary that addresses a more contemporary set of goals. In the early twenty-first century, when many in the industrialized world feel overwhelmed by human artifice and dream of escaping into a primeval wilderness, Cameron's parable of Jake Sully caters to that fantasy, but with its own paradoxical caveat: Technology, to be escaped, must be embraced completely.

If Cameron's *Avatar* speaks to the enduring nostalgia for a lost Eden, Christopher Nolan's *Interstellar* speaks to another dream that has its deep roots in the Abrahamic religious traditions: the escape from a troubled and corrupted earth into a celestial paradise. In Christianity and Islam especially, the dream of a better life in an eternal world above and beyond this one has exercised a powerful influence. Although visions of Eden and the Heaven beyond the stars are similar in key respects, each speaks to a distinct set of aspirations. If the return to Eden is a trope for a restored sense of unity with nature, the dream of ascending to heaven signals the complete restoration of our unity with a celestial God. Although *Interstellar* is a work of science fiction that makes no direct allusion to traditional religious themes, it still follows the broad contours of a tale of faith in which the protagonist rejects sin, endures a series of trials, and then attains an eternal state of redemption that transcends death. In the case of *Interstellar*, however, that faith is not placed in an anthropomorphic deity, but in the technological ability of the human race to escape the confines of the earth and discover a better life in outer space.

The hero who maintains his faith in that premise is a former test pilot named Cooper. In a world darkened by a global crop-destroying blight and terrifying dust storms, Cooper maintains his faith in the future, and shares a powerful love for science and space exploration with his daughter Murphy. Early in the film, Cooper faces down a schoolteacher who has embraced a new public-school curriculum which denies that the Apollo missions to the moon ever happened, while asserting that the current generation of humans are merely "caretakers" who must maintain the low-level production of corn (the only crop that still resists the global blight), and pass on a greatly diminished but still marginally livable world to the next generation. His faith in the future is tested by this unpleasant encounter, but Cooper remains defiant and continues teaching his daughter about science at home (*Interstellar* 2014).

Cooper's lonely commitment to his faith is rewarded when he stumbles upon a government program that has been kept secret from the public. In a stunning turn of events, he learns that NASA is preparing an interstellar mission for discovering another habitable planet for the people of Earth, and they want him to be the lead pilot. The plot of *Interstellar* then unfolds with trials and tribulations worthy of the Puritan allegory, *A Pilgrim's Progress*. After numerous adventures with interstellar spaceflight and the time-bending effects of special relativity, Cooper must fight for his life when he encounters a rogue astronaut who has committed the ultimate sin in science, falsifying data to serve his own personal interests. Cooper prevails in that battle and then achieves a series of heroic feats that enable him to save the NASA mission from complete destruction. Ultimately, he passes through a black hole and finds himself in an extra-dimensional "tesseract" where time is represented as a spatial construct. From this place outside of time, he is able to communicate his love to his daughter Murphy on Earth, and succeeds in spurring her to take steps that will help her to save not only the NASA mission, but also the human race.

When Cooper returns from his interstellar travels to our solar system, decades have passed. Though he has barely aged at all, his daughter is now an old woman, surrounded by her children and grandchildren as she lies dying in a hospital bed. The inescapable sadness of her growing old before her father is more than mitigated, however, by the tremendous success of the technological project that both of them believed in and protected. Human civilization has completely left the earth behind and now thrives in a gigantic space colony somewhere close to Saturn. Modeled on the space colonies first proposed in the 1970s by Gerard K. O'Neill, this rotating tubular colony has everything from artificial gravity, to suburban streets, to scenes of baseball games that combine the familiar Americana of Norman Rockwell with the unnervingly paradoxical perspective of M. C. Escher. In the success of this colony, the faith of the senior NASA engineer who chose Cooper to lead his mission is affirmed: "We're not meant to save the world. We're meant to leave it, and this is the mission you were trained for." Because of his dogged determination, his faith in the efficacy of science and technology, and his timeless love for his daughter, Murphy was able to carry out his mission, and the human race was able to leave a dying earth behind and build a new world for itself among the stars. The ancient religious dream of escaping nature and finding a better home in the heavens is secularized in *Interstellar*, but its broad contours remain the same.

As popular films, both *Avatar* and *Interstellar* are compelling entertainment, and each owes its success, not only to its highly immersive special effects, but also to its strong characters and suspenseful storylines. As narratives that explore the relationship between technology and nature, both films present an odd combination of powerful ideas and irreducible paradoxes. On one level, *Avatar* reads like an indictment of our technologically driven culture of consumerism and its rampant consumption of natural resources. On the other hand, the solution that it offers to its protagonist, complete transformation into one of the Na'vi, is one that would be utterly inconceivable without the very panoply of manipulative technologies that *Avatar* frames as dangerously exploitative. Taking a very different tack, *Interstellar* exalts science and technology and seems to posit that the greatest danger for the human race is not posed by our overexploitation of resources, but by a return to the medieval mindset of fatalism that preceded the scientific revolution. But this vision is dogged by its own contradictions. The premise that it is the destiny of humankind to leave the earth rather than save it ignores the crucial fact that we have never thrived – or even survived – as a species without a vast ecological infrastructure comprising numerous other plant and animal species great and small. If we could not solve the crop-killing blight that threatens the human race on earth, there is no chance that we would be able to solve it on a space colony orbiting the planet Saturn. Just as the aspiring Na'vi warrior Jake Sully in *Avatar* cannot dispense with advanced technology in his quest to become one with nature, the intrepid technophiles of *Interstellar* cannot dispense with the life-sustaining services of countless other species, no matter how clever they are at building new kinds of orbiting suburban subdivisions or blasting their way through black holes. Each film depicts an escape from the human predicament that is every bit as impossible as it is appealing. For our species, the marriage between technology and nature may or may not have a happy future, but one thing is certain: divorce is impossible.

The intellectual dichotomy between human rights and ecological sustainability that characterizes the work of Malthusian environmentalists such as Garrett Hardin, and the cultural dichotomy between technology and nature that permeates so many popular visions of the future, have each done their part to distort our understanding of the environmental challenges of the Anthropocene. In addition to these, an irrational reverence for the concept of national sovereignty has resulted in a third kind of false dichotomy that infects our political thinking. When the people of any nation define their "national security" as distinct from that of every other nation on earth, they become blind to those problems, such as climate change, that present a serious threat to the security of us all. If we are to embrace clear thinking and concerted action to address the problem of climate change, we must see past the political divisions of the Westphalian system of nation states, just as we must see past such intellectual and cultural divisions between humanity and nature. The work of Elisabeth Mann Borgese, which will be considered in the next chapter, transcended all of these false dichotomies.

## Bibliography

Almond, Philip C. 1999. *Adam and eve in seventeenth century thought*. Cambridge: Cambridge University Press.

Anderson, Benedict. 1983. *Imagined communities: Reflections on the origin and spread of nationalism*. London: Verso.

Borgese, Elisabeth Mann. 1965. *A constitution for the world*. Santa Barbara: Center for the Study of Democratic Institutions.

Burke, Edmund.[1790] 2012. Reflections *on the* revolution *in France*. Mineola: Dover Publications.

Cameron, James and Jon Landau (producers) and Cameron, James (Director). 2009. Avatar. United States: Twentieth Century Fox.

Cloud, Preston. 1989. *Oasis in space: Earth history from the beginning*. New York: W. W. Norton & Co.

Collier, Aine. 2007. *The humble little condom: A history*. New York: Prometheus Books.

Deese, R.S. 2008. A metaphor at midlife: 'The tragedy of the commons' turns 40. *Endeavour* 32 (4): 152–155.

———. 2009. The artifact of nature: Spaceship earth and the dawn of global environmentalism. *Endeavour* 33 (2): 70–75.

———. 2015. *We are amphibians: Julian and Aldous Huxley on the future of our species*. Oakland: University of California Press.

Evans, James. 1998. *The history and practice of ancient astronomy*. Oxford: Oxford University Press.

Garan, Ron. 2015. *The orbital perspective*. Oakland: Berrett-Koehler Publishers.

Hardin, Garrett. 1968. The tragedy of the commons. In *Managing the commons*, ed. Garrett Hardin and John Baden. San Francisco: W.H. Freeman and Company. (1977), 20, 24.

———. 1973. *Exploring new ethics for survival: The voyage of the spaceship beagle*. London: Penguin Books.

———. 1993. *Living within limits: Ecology, economics, and population taboos*. Vol. 21, 24–25. Oxford: Oxford University Press.

Hardin, G. 1999. The persistence of the species. *Politics and the Life Sciences* 18 (2): 225–227. Retrieved from http://www.jstor.org.ezproxy.bu.edu/stable/4236506.

Hesiod. 2017. *The poems of hesiod: Theogony, works and days, and the shield of herakles* Trans by Barry B. Powell. Oakland: University of California Press.

Hill, James. 2011. *Descartes and the doubting mind*. London: Bloomsbury.

Holden, Barry. 2002. *Democracy and global warming*, 6–7. London: Continuum.

Howe, Nicolas. 2016. *Landscapes of the secular: Law, religion, and American sacred space*. Chicago: University of Chicago Press.

Huxley, Aldous. [1962] 2002. *Island*. New York: Harper Perennial Classics.

Josephson, Paul R. 2005. *Resources under regimes: Technology, environment, and the state*. Cambridge: Harvard University Press.

Keohane, Robert O., and Elinor Ostrom. 1995. *Local commons and global interdependence: Heterogeneity and cooperation in two domains*. London: Sage Publicaitons.

Lovelock, James. 1995. *The ages of Gaia: A biography of our living earth*. Oxford: Oxford University Press.

MacAskill, Ewen 2008. "Hurricane Gustav: Republican Convention Thrown into Chaos" *The Guardian*, 31 August, 2008. https://www.theguardian.com/world/2008/sep/01/usa.republicans2008

McKibben, Bill. 2010. *Eaarth: Making life on a tough new planet*. New York: Henry Holt and Company.

Monnet, Jean., Richard Mayne., trans. 1978. Memoirs. New York: Doubleday & Company.

Nolan, Christopher with Lynda Obst and Emma Thomas (Producers) & Nolan, Christopher 2014. Interstellar. United States: Paramount Pictures.

Ostrom, Elinor. 2009. "*Beyond markets and states: Polycentric governance of complex economic systems*" Nobel Prize Lecture, December 8, 2009 https://www.nobelprize.org/nobel_prizes/economic-sciences/laureates/2009/ostrom_lecture.pdf. (Accessed 11 Jan 2018)

Poole, Robert. 2008. *Earthrise: How man first saw the earth*. New Haven: Yale University Press.

Reves, Emery. 1942. *A democratic manifesto*. New York: Random House.

Ricks, Thomas. 2017. *Churchill and Orwell: The fight for freedom*. New York: Penguin Press.

Robertson, Thomas. 2012. *The malthusian moment: global population growth and the birth of american environmentalism*. New Brunswick: Rutgers University Press.

Ruse, Michael. 2013. *The Gaia hypothesis: Science on a pagan planet*. Chicago: University of Chicago Press.

Sanderson, Warren C. 2004. *The end of world population growth in the 21st century: New challenges for human capital formation and sustainable development*. London: Routledge.

Southern Poverty Law Center (SPLC). 2017. https://www.splcenter.org/fighting-hate/extremist-files/individual/garrett-hardin (Accessed 11 Aug 2017).

Stanton, Elizabeth Cady. 1848. (2015). A declaration of sentiments and resolutions. In *Carlisle*. Massachusetts: Applewood Books.

*The Social Contract*, 1997. *Interview with Garrett Hardin*. Vol. 8, No. 1. Fall 1997. Petoskey: The Social Contract Press.

Vandevelder, Paul. 2009. *Coyote warrior*. Nebraska: University of Nebraska Press.

Vonnegut, Kurt. 1974. *Wampeters, Foma & Granfalloons*. New York: Delacorte Press.

# Chapter 5
# Transcending the Tragedy of the Commons

*"The Earth provides enough to satisfy every man's need but not for every man's greed."*

Mahatma Gandhi

**Abstract** The work of Elisabeth Mann Borgese represents an important bridge between the world federalist and ocean conservation movements in the twentieth century, and offers a viable path for overcoming the false dichotomy between human rights inherent in much of our discourse about the Tragedy of the Commons. Her vision of universal human rights and the protection of the oceans as the "common heritage of humanity" helped to shape the third UN Convention on the Law of the Sea (UNCLOS) in 1982, and articulated principles that could be useful in establishing the global rule of law in order to protect the atmosphere and climate of the Earth in the twenty-first century.

**Keywords** Elisabeth Mann Borgese · World federalism · Law of the sea · Common heritage

As we have seen, the false dichotomy between human rights and ecological sustainability has pervaded the rhetoric of neo-Malthusian environmentalism, and the false dichotomy between technology and nature has pervaded our popular culture. The concept that natural places should be preserved as "pristine wilderness" and untouched by human influence is an appealing one, but it has become both untenable and counterproductive for a number of reasons. It has become untenable because, in the Anthropocene, there is literally no place on earth that has not been altered by human activity. It has become counterproductive because it ignores innumerable natural resources, such as city parks and green spaces, which do not fit the traditional definition of wilderness.

In his essay, "The Trouble with Wilderness; or, Getting Back to the Wrong Nature," the environmental historian William Cronon argued that this dichotomy



R. S. Deese, *Climate Change and the Future of Democracy*, Environmental Challenges and Solutions 5, https://doi.org/10.1007/978-3-319-98307-3_5

has led many environmentalists to undermine their own efforts to advance sustainability:

> Idealizing a distant wilderness too often means not idealizing the environment in which we actually live, the landscape that for better or worse we call home. Most of our most serious environmental problems start right here, at home, and if we are to solve those problems, we need an environmental ethic that will tell us as much about using nature as about not using it. The wilderness dualism tends to cast any use as abuse, and thereby denies us a middle ground in which responsible use and non-use might attain some kind of balanced, sustainable relationship. My own belief is that only by exploring this middle ground will we learn ways of imagining a better world for all of us: humans and nonhumans, rich people and poor, women and men, First Worlders and Third Worlders, white folks and people of color, consumers and producers—a world better for humanity in all of its diversity and for all the rest of nature too. The middle ground is where we actually live. It is where we—all of us, in our different places and ways—make our homes. (Cronon 1996, p. 86)

Cronon's seminal essay has inspired many to rethink the values that have driven campaigns for environmental causes. Efforts to define and protect the "middle ground" that Cronon describes here have been a major factor in the movement for sustainable development over the past two decades.

Elisabeth Mann Borgese, whose career as a global activist began soon after World War Two and extended into the twenty-first century, was one writer and activist whose work anticipated Cronon's vision of the "middle ground" by several decades. She possessed a unique ability to see through false dichotomies and advance the cause of sustainable development on a global scale. If we trace her career from her work on the Committee to Frame a World Constitution at the University of Chicago in the 1940s to her work on ocean conservation between 1970 and 2002, her ideas about the global commons and how to preserve it emerge as distinctly perceptive and applicable to the problem of climate change. After the destruction of Hiroshima and Nagasaki in August of 1945, the president of the University of Chicago organized a committee to draft a world constitution, and Elisabeth Mann Borgese would soon become the most articulate advocate for this project. Hutchins made no secret of the shock that he felt as the enormous dangers posed nuclear weapons came into focus. Less than three years before, his university had agreed to participate in the Manhattan Project and had been the site of the first controlled nuclear reaction, supervised by Enrico Fermi. Now, the advances in nuclear research made in Chicago had led to the destruction of two Japanese cities, signaling the arrival of an era in which total war between sovereign nation states meant total destruction for human civilization. Feeling an acute responsibility for what had happened, Hutchins convened a group of scholars from across the disciplines to investigate how a new framework for world government might be created to prevent another world war. Among academics, Robert Maynard Hutchins had distinguished himself as one who attempted not only to study problems, but to look for solutions. During the first half of the twentieth century, the forces unleashed by two world wars convinced many scientists and intellectuals that our species would not survive unless new supranational institutions and norms were created to check the power of ambitious nation states in order to preserve peace.

During the second half of the twentieth century, the challenge of ecological disintegration on a global scale, accelerated by climate change, convinced later generations that more democratic, responsive, and effective supranational institutions would be required, not only to prevent war, but also to protect the biosphere. The life and work of Elisabeth Mann Borgese (1918–2002) represents a remarkable bridge between the world federalist movement that thrived immediately after World War Two and the global environmentalist movement that began to gather force in the 1960s and 1970s. Her contribution to the creation of the third UN Convention on the Law of the Sea (UNCLOS), which was opened for signature in 1982 and has been accepted by the vast majority of the world's governments since 1994, reflects her lifelong commitment to the values of democracy, peace, and ecological sustainability. In a century characterized be rampant intellectual specialization and ideological sectarianism, Borgese's work transcended such barriers and was fired by her conviction that peace could not be had without democracy, and ecological sustainability could not be had without peace.

Elisabeth Mann Borgese's vision of expanding democracy to address global ecological crises was well founded. In assessing the value of democratic institutions for dealing with complex problems such as climate change, the political science scholar Barry Holden points to "Aristotle's argument that knowledge is scattered throughout the people as a whole; and the range of persons this brings in must imply a range of knowledge: 'although each individual may be deficient in the qualities necessary for political decision-making, the people collectively are not deficient in this way: they are indeed better endowed than any experts, since the *combined* qualities of the individuals add up to a far from deficient totality'" (Holden 2002, pp. 195–196). While she recognized the unique adaptability and resilience of democracy, Elisabeth Mann Borgese was aware that "Western democratic theory, from the days of Mill to our own, has been concerned with domestic affairs" (1965, p. 19). In the age of accelerating globalization that characterized the second half of the twentieth century, matters of "foreign policy" that had previously been beyond the purview of democratic politics, have begun to have had an increasingly dramatic impact on domestic affairs:

> When foreign affairs, the crushing issues of war and peace in a technologically shrinking, increasingly interdependent world, begin to outweigh internal affairs, and domestic policy becomes largely determine by foreign policy, the democratic process is doomed to be stifled and choked. To be rescued, to survive even on the domestic plane, the democratic process must be carried over from the internal to the international sector (Borgese 1965, pp. 19–20).

In other words, if we want democracy to survive in an age of globalization, we must create new democratic institutions so that global affairs will be conducted "not by diplomats representing the executive, but by representatives of the people in international bodies of deliberation" (Borgese 1965, p. 20).

Elisabeth Mann Borgese was perhaps especially suited to examine the power of democratic ideals in the face of global crises. Born in 1918 to the German writer Thomas Mann and his wife Katia Pringsheim Mann, she came of age on a continent shadowed both by total war and by the rise of totalitarian government, before her family moved to the United States in 1938. Committed to the prevention of war through the expansion of democratic freedom and accountability across national boundaries, she became involved in the Committee to Draft a World Constitution at the University of Chicago immediately after World War Two. Although the world federalist movement was greatly weakened by the Cold War polarization of the 1950s, Elisabeth Mann Borgese would not abandon her vision of building a framework for intelligent cooperation and democratic accountability on a global scale. In the late 1960s, she worked with former University of Chicago President Robert Maynard Hutchins at the Center for the Study of Democratic Institutions in Santa Barbara to advance the cause of ocean conservation. In 1970, she helped conceive of and convene the first *Pacem in Maribus* (Peace in the Oceans) Conference in Malta. This conference spurred the creation of the International Ocean Institute, an NGO dedicated to ocean conservation founded in 1972. In the realm of international law, the *Pacem in Maribus* Conference accelerated the painstaking process of advancing negotiations for UNCLOS, which was opened for signature in 1982. Her chief contribution to this treaty was to affirm the principle, first articulated by the Committee to Frame a World Constitution at the University of Chicago in the late 1940s, that any universal commons, such as the seabed under international waters, must be protected as the common heritage of the entire human race.

As the youngest daughter of Thomas Mann and Katia Pringsheim Mann, Elisabeth grew up in an atmosphere where political cosmopolitanism was part of her intellectual inheritance. Her father was an outspoken opponent of ultranationalism in Germany for years before the rise of Hitler, and her mother was the granddaughter of one of the most outspoken European feminist authors of the nineteenth century, Hedwig Dohm. One of Dohm's most famous statements ("Die Menschenrecht habt keine Geschlecht" or "Human rights have no gender") appears as the epitaph on her gravestone, and was a credo that Elisabeth exemplified with her life's work.

Although Elisabeth's father was perhaps the most famous German author of the twentieth century, Thomas Mann was a better exemplar of Enlightenment cosmopolitanism than he was of any particular national culture. The novel for which he won the Nobel Prize, *Buddenbrooks*, presents itself as the chronicle of a merchant family in decline, but it can also be read as a lamentation for the loss of a pre-national culture in German seafaring cities such as Lübeck that were more economically and culturally connected to the seafaring and cosmopolitan Hanseatic League than to the more provincial and militaristic hinterlands of what would become the German nation. In *Buddenbrooks*, the sea itself is a symbol for freedom, equality, and genuine love. When a young Antonia Buddenbrook asks her beloved, the son of a sea captain who has joined a secret liberal society at his university, to explain what he is struggling for, he gestures to the North Sea:

> "We want freedom," Morten said.
>
> "Freedom?" she asked.

> "Yes, freedom, you know – *Freedom*!" he repeated; and he made a vague, awkward, fervent gesture outward and downward, not toward the side where the coast of Mecklenburg narrowed the bay, but in the direction of the open sea, whose rippling blue, green, yellow, and grey stripes rolled as far as eye could see out to the misty horizon.
>
> Tony followed his gesture with her eye; they sat, their hands lying close together on the bench, and looked into the distance. Thus they remained in silence a long time, while the sea sent up to them its soft enchanting whispers. . . . Tony suddenly felt herself one with Morten in a great, vague yearning comprehension of this portentous something which he called "Freedom." (Mann 1901 1983, p. 115).

When Antonie is compelled by her father to marry a corrupt businessman whom she finds unctuous and revolting in order to bolster the financial position of the Buddenbrook family, she must turn her back on a marriage based on elective affinities, and on the idea of freedom that she had attempted to grasp on the afternoon by the sea. Years later, after the mercenary marriage that her family compelled her to accept has ended in disaster, and her second marriage to a provincial Bavarian has met a similar fate, Antonie remembers the sea with a deeps sense of loss that she struggles to articulate.

Thomas Mann's model for Antonie Buddenbrook was a beloved aunt named Elisabeth Mann who died in 1917. The following year, when his youngest daughter was born, she was given the same name. Thomas Mann had no scruples to prevent him from declaring often and quite publicly that she was his favorite child. Like Antonie Buddenbrook, some of young Elisabeth's most formative memories involved the sea, and she came to see it as a powerful symbol for freedom. As a very young girl, Elisabeth saw her family reprimanded by local authorities of the Italian Fascist government for letting her bathe naked in the Mediterranean. Committed to the cause of Antifascism from an early age, Elisabeth married the outspoken Sicilian Antifascist Giuseppe Antonio Borgese in 1939, on the eve of the Second World War. In her first writings on the legal status of the sea in the late 1960s, she wrote in language that echoed *Buddenbrooks*:

> The Oceans are free. The mere thought that they could be "appropriated" by any ruler however mighty, by any nation, no matter how vast its empire, has something blasphemous. The oceans, in a way, are the most sublime expression on earth of what is extra-human, superhuman, indomitable. That the oceans are free is, in fact, the oldest of all international laws. Back in the sixteenth century, Ivan the Terrible called the ocean "God's Road," and Queen Elizabeth I of England, in disposing of the Spanish Ambassadors' complaints on the depredations by Sir Francis Drake on the Spanish treasure fleet, is quoted as having said: "The use of the sea and the air is common to all. Neither can title to the oceans belong to any people or private person forasmuch as neither nature nor public use or custom permitted any possession thereof" (Borgese 1969).

It must be conceded that freedom is an abstract concept, while the sea is an immense and tangible reality. We perceive an abstraction with our intellect and a tangible reality with our senses. The rare genius of a great artist is to bring these two modes of perception together so that what is meaningful to our intellect may become tangible to our senses, and what it is tangible to our senses may become meaningful to our intellect. For her own part, Elisabeth Mann Borgese did not care to think of genius as an inherited trait (Bender 1969. p. 50). Whether she inherited this gift from her illustrious father, or acquired it from her upbringing, it was clearly an ability that both of them shared. As one of the leading literati of the twentieth century, Thomas Mann was expected to have such skills, but in the field of international relations and political science, where Elisabeth Mann Borgese made her reputation, these skills are exceedingly rare. In a world of dry and colorless prose, larded with technical jargon and an alphabet soup of acronyms, Elisabeth Mann Borgese was the rare combination of an expert and a poet.

The single most important idea that Borgese brought to the issue of ocean conservation was the principle that the seabed under international waters must be legally established as the common heritage of the entire human race. Her role in developing and promoting this idea began with her experience as part of the Committee to Draft a World Constitution at the University of Chicago immediately after World War Two. As part of this project, she collaborated with university president Robert Maynard Hutchins, the philosopher Richard McKeon, and with her husband Giuseppe Antonio Borgese. Drawing on the work of dozens of scholars, the committee quietly studied extant and proposed democratic constitutions from around the world as it attempted to design a framework for the stable and sustainable practice of representative democracy on a global scale (Baratta 2004a, b. Vol. 1, p. 218).

Its draft constitution was published in 1948. The fact that newspapers in both the Soviet Union and the United States forcefully condemned this project reflects the determination of the Chicago committee to transcend the limitations of both doctrinaire ideology and narrow nationalism. Jealously protecting Stalin's conception of Soviet national interest, *Pravda* condemned the Committee to Draft a World Constitution as a Trojan horse for American imperialism. Meanwhile, conservative American papers such as the *Chicago Tribune* aired suspicions that the Committee was animated by a desire to advance Marxist communism, warning its readers that a "super-secret constitution" had been cooked up by "one of a rash of militant globalist organizations which have sprung up in the United States and England since the United Nations has demonstrated its uselessness." The *Tribune* surmised that the document "appears to be a combination of Franklin D. Roosevelt and Karl Marx." From the left, condemnation was no less swift. Moscow Radio quoted Soviet journalists who saw the project as a capitalist ruse, a clever instrument "to justify the American Empire plan for world supremacy," warning that "the program of the Chicago world government embodies the ambitions of the American warmongers" (Boyer 1995). Of course, since this project was being conducted in the United States, the charge that it was another capitalist plot may have had some credibility for *Pravda* readers. Likewise, the charge that the Committee to Draft a World Constitution was establishing a beachhead in Chicago for the advance of world communism could point to a key concept in the document that it produced. This was the principle that the classical elements essential to life, i.e. earth, water, air, and energy, must be legally recognized as "the common property of the human race" (Boyer 1995).

Though the vigilant editors of the *Chicago Tribune* condemned this principle in the late 1940s as a Marxist ruse, it had much deeper roots in the history of human civilization. As the historian Peter Linbaugh has documented in his 2008 book *The Magna Carta Manifesto*, the foundational *Magna Carta* of 1215 drew much of its power from another document promulgated at the same time, the *Charter of the Forest*. This charter guaranteed ordinary people access to common lands for gathering food, grazing cattle, hunting, and gathering firewood. Of course, in their allusion to the classical elements of earth, air, fire, and water, the drafters of the World Constitution reached even further back, to the philosophical and political discourse of the ancient Greeks.

Elisabeth Mann Borgese passionately advocated for the inclusion of this principle in the third UN Convention on the Law of the Sea by arguing that the seabed under international waters was the common heritage of the entire human race. This principle, though more narrowly applied than Borgese would have liked, was enshrined in the Law of the Sea when it was drafted in 1982. It led to the creation of the International Seabed Authority (ISA), based in Kingston, Jamaica. Under the UN Convention on the Law of the Sea, mining projects of mineral resources on the ocean floor under international waters are regulated by the ISA and a portion of the profits from this economic activity go into a common resource fund that is available to aid economic development in all nations, including those which are landlocked.

After the negotiations for UNCLOS were completed and it was opened for signature in 1982, Arvid Pardo, the UN ambassador from Malta and a close confidant to Elisabeth Mann Borgese, expressed some disappointment that the treaty did not go further in safeguarding natural resources and establishing the rule of law on a global scale. On balance, UNCLOS leaves ample leeway for the territorial claims of nation states and the commercial interests of multinational corporations. The concept of the Exclusive Economic Zone (EEZ), allows many seafaring nations to assert their sovereign interests further from their shores than ever before. From the point of view of multinational corporations involved in major industries such as fossil fuel extraction, shipping, and fishing, UNCLOS does little to curtail their operations. The oil industry, for example, can still conduct extensive offshore drilling on the continental shelf of any nation that approves those activities, a distance that UNCLOS recognizes as stretching 200 nautical miles out to sea from the coast of any sovereign nation. In spite of the fact that UNCLOS is actually quite accommodating to national sovereignty and to commercial interests, the principle that the ocean floor under international waters must remain the common property of the entire human race has proven unacceptable in some quarters. In particular, advocates of traditional nationalism and free market fundamentalism at organizations such as The Heritage Foundation and the Cato Institute see this concept as a poison pill. For this reason, the Reagan administration refused to sign UNCLOS in 1982, while former Nixon speechwriters Pat Buchanan and William Safire advanced a broad critique of the treaty in the press, even dusting off the charge, leveled at Elisabeth Mann Borgese and her Chicago colleagues in the late 1940s, that the principle of sharing the wealth of a global resource must be some sort of Marxist plot. Building on the old chestnut of creeping Marxism, Safire added a racially charged metaphor when he sounded the alarm that, "the third world camel has its nose in the tent" when he urged the Reagan administration to kill the treaty (Safire 1982).

In an attempt to stop acceptance of the treaty among Western nations, the Reagan administration sent former (and future) Secretary of Defense Donald Rumsfeld as a special envoy to the United Kingdom, West Germany, and France to propose an alternative framework for exploiting mineral resources under international waters entirely outside the framework of the Law of the Sea treaty (Safire 1982). Given the size of these economies, a success for Rumsfeld in this endeavor would have effectively destroyed the treaty. However, not one of the governments he approached was willing to abandon the Law of the Sea treaty. Rumsfeld's attempt to create what we might call a "coalition of the unwilling" failed, largely because the treaty as it stood was fairly accommodating to national and corporate interests. By 1994, more than 160 nations had endorsed UNCLOS, thus endowing it with the status of accepted international law. The Clinton administration agreed to sign the treaty, but did not submit it to the Senate for ratification. To date, the United States still has not ratified UNCLOS, though the U.S. Navy has declared that it will accept it as law. The principle that Elisabeth Mann Borgese helped to articulate at the University of Chicago – the concept of the common heritage of the human race – has proven to be too radical for American opponents of UNCLOS.

The specific issue that inspired the creation of the International Seabed Authority was the desire to mine manganese nodules from the surface of the seabed below international waters. These nodules, pieces of igneous rock that resemble burnt potatoes, are rich in valuable minerals such as copper, nickel, and cobalt. Though the market price of these minerals was rapidly rising in the 1960s and 1970s, it has since stabilized to an extent that the cost of harvesting manganese nodules from the ocean floor still outstrips their market value. Thus, the refusal to accept the International Seabed Authority is in reality the refusal to accept the diminution of *potential* profits in a *theoretical* future rather than the loss of actual profits in the present. For this reason, the vast majority of countries, including nations such as Japan with extensive maritime interests, have been willing to accept UNCLOS. Those who have been resolutely opposed, such as the Reagan administration, the U.S. Senate, and think tanks such as the Heritage Foundation, have been opposed for strictly ideological rather than economic reasons. They have recognized that protecting the seabed under international waters as the "common heritage of mankind" would establish a powerful principle that could limit the activities of private industry in other common areas, such as the arctic, Antarctica, and even outer space (Jakhu et al. 2017. p. 381). Most important, this principle would establish a precedent for limiting the actions of private industry in order to protect the common resources of the earth's atmosphere and climate.

Almost two decades after the UN Convention on the Law of the Sea was first opened for signing, Elisabeth Mann Borgese reflected upon how it had both challenged and changed received notions of sovereignty and helped to advance the evolution of a "global civil society." She added that, "one might even go so far as to say" that the provision that UNCLOS makes for protecting the global commons "makes 'mankind' a subject of international law—the ultimate transcendence of the sovereign nation state" (Borgese 1999, p. 986). Because the sea defied the imposition of static boundaries, maritime law represented the ideal arena for crafting transnational solutions for problems that affect the entire planet. Borgese concluded that the "oceans are a great laboratory for the fashioning of a sustainable order, in which an increasingly global civil society plays its leading role in the Global Century" (1999, p. 991). In spite of the resistance her efforts had faced, especially from the government of the United States, Borgese remained convinced that time and tide were both tipping the balance away from the fetishization of national sovereignty and toward more effective global cooperation to protect our oceans, atmosphere, and biosphere. The essential prerequisite to such cooperation was, as Borgese saw it, "the democratization of global governance" (1990. p. 990).

In the seventeenth century, Hugo Grotius made the argument for *Mare Liberum*, or a free ocean, based on the premise that it is not divisible in the way that land is. In the twenty-first century, we are discovering something that Elisabeth Mann Borgese bore witness to for her entire career as a writer and activist. Her work in ocean conservation was part of a broad effort in the late twentieth century that proved to be remarkably effective. As the historian Mark Mazower chronicles, intergovernmental cooperation to preserve the oceans made tremendous strides in the 1970s, when "as many treaties were drawn up in a single decade as in the previous forty years" with the consequence "that the seas became markedly cleaner as agreements were reached to control marine dumping and land-based sources of marine pollution as well" (Mazower 2012, p. 336). The Mediterranean, where pollution had reached a "critical level" by the late 1960s, benefited from agreements in the subsequent decade to limit marine dumping, as well as sources of contamination from industry and agriculture in the nations that surrounded it. Thanks to the steady coordination of the United Nations Environmental Program (UNEP), "pollution levels were stabilized, and water cleanliness improved despite the rapid growth of cities and industries around its shores" (Mazower 2012, p. 336).

This progress was the result of a collaboration between environmentalists and "One Worlders" such as Elisabeth Mann Borgese. These individuals realized that the common heritage of the human race is essential to all of our lives, and it will be destroyed if it is conceived of as divisible. If the oceans could be divided, owned, and degraded in the name of private enterprise, so could the atmosphere that we breathe and the climate on which we depend as the lynchpin of our ecosystems. If we destroy the oceans in the name of private property and national sovereignty, we will destroy ourselves. Elisabeth Mann Borgese dedicated the last three decades of her life to this realization: Only if we save the oceans for the sake of our common interests, can we manage, as a species, to save ourselves.

This link between the protection of the global commons and the fate of our species was becoming increasingly apparent in the 1970s. In 1972, the United Nations sponsored an international conference in Stockholm, Sweden to address threats to the biosphere and to enhance international cooperation to deal with those threats. To sum up the findings of the Stockholm Conference on the Human Environment, the economist Barbara Ward and the microbiologist René Dubos collaborated on a book entitled *Only One Earth*. In their conclusion, they lamented that, "The planet is not yet a center of rational loyalty for all mankind" (Ward and Dubos 1972, p. 220.) They anticipated, however, that it would be possible for human beings to cultivate and strengthen such a planetary loyalty, without betraying the time-honored loyalties of filial piety and patriotism:

> From family to clan, from clan to nation, from nation to federation – such enlargements of allegiance have occurred without wiping out the earlier loves. Today, in human society, we can perhaps hope to survive in all our prized diversity provided we can achieve an ultimate loyalty to our single, beautiful, and vulnerable planet earth. (Ward and Dubos 1972, p. 220)

In the early twenty-first century, it remains to be seen whether this sense of planetary loyalty is any more likely to take hold than it was in 1972. However, there is reason to believe the claim of Ward and Dubos that new and broader loyalties need not displace older and more localized loyalties. The expansion of allegiance from family to nation is an excellent example of this. For much of human history, it was commonplace for individuals in one family to reserve the right to use lethal force against members of another family. Longstanding feuds, whether between the houses of Capulet and Montague or Hatfield and McCoy, were the stuff, not only of literature, but of history as well. Such feuds became forbidden with the rise of the modern state, which according to Max Weber's definition, "lays claim to the *monopoly of legitimate physical violence* within a particular territory" (Weber, Max. [1919] 2004, p. 33). Few reasonable people would argue, however, that a family which no longer retains the right to carry on blood feuds with rival families is less loyal, loving, and cohesive than a family that retains such a right. In fact, as countless dramas from *Romeo and Juliet* to *The Godfather* would suggest, families that are caught up in blood feuds with their rivals are likely to be riven by internal conflicts that can be just as lethal. The same logic applies to the nation state. As members of the European Union, France and Germany have set aside the prerogative of going to war with each other as a means of settling their differences. This does not make their citizens less patriotic than, say, the citizens of India and Pakistan, for whom the possibility of cross border conflict remains constant. Just as we can love our own family without reserving the right to feud with another family, we can love our own nation without reserving the right to wage war against another nation. In their coda to *Only One Earth*, Ward and Dubos describe the steady expansion of loyalty as an expansion of consciousness.

Of course, it remains apposite to remember Weber's point that the authority of the state is founded on its hard power; it is legitimate because it retains a monopoly on the use of force within the territory under its control. As recent examples of civil unrest, such as the Los Angeles riots of 1992 or the aftermath of Hurricane Katrina in 2005 have indicated, families and other social groups will quickly resort to violence to protect their interests if they perceive that law enforcement is either absent or unreliable. Likewise, an increased sense of anarchy in international relations will increase the likelihood that nation states, regardless of their previous commitments to an international order based on law, will resort to war or the threat of war to protect their interests.

There is also little room for doubt that competition among sovereign political entities has been a spur to exploration and innovation. When historians attempt to explain the rapid rise in technological innovation that gathered force in Europe after 1500, the element of competition among sovereign nation states is often cited as a factor. Not only did such competition spur technological innovation and nautical exploration, but it also prevented the dominance of a single overarching state authority that might become an obstacle to such innovation and exploration. In this respect, fifteenth-century China under the rule of the Ming Dynasty provides an instructive contrast to Renaissance Europe during the same period. At the beginning of the fifteenth century, the Ming Dynasty sponsored the naval expeditions of the admiral Zheng He, who sailed a fleet of massive "treasure ships" to Africa and the Arabian Peninsula. Zheng's ships were very well constructed and many times larger than the ships later sailed by Columbus to the Western Hemisphere. Zheng's achievements placed China on the cusp of even greater achievements in the field of nautical exploration, but his program was cut short when a new emperor came to power in 1424. Because China was ruled by a single monarchy and bureaucracy, it was possible for that single state authority to close the door on further maritime voyages decisively. In Europe, by contrast, a loose system of competing polities could afford an ambitious explorer such as Christopher Columbus the opportunity to propose his idea for an expedition to Asia to several monarchies, including Portugal and England, until he found sponsorship from Ferdinand and Isabella of Spain (Diamond 1999, p. 412). Nearly five centuries later, competition between sovereign states continued to drive innovation and exploration, as evidenced by the space race between the United States and the Soviet Union.

However, such competition need not be military, nor even strategic in its applications. International competition in athletic venues such as the Olympics or World Cup has become a global obsession, and indicates that competition in areas above and beyond military prowess can capture the popular imagination of citizens around the world. Furthermore, the significant attention garnered by the Nobel Prize awards each year suggests that the excitement of international competition for achievement and recognition is not limited to the field of sports. If other globally recognized prizes could be awarded to researchers and inventers in such fields as renewable energy, sustainable agriculture, and disaster relief, it is not unreasonable to expect that they would stimulate new levels of innovation reminiscent of the space race during the 1960s. If we divorce the concept of national prestige from military might, we will be able to reap the benefits of competition without the specters of war and nuclear holocaust that shadowed the technological competition between the United States and the Soviet Union in the age of Kennedy and Khrushchev. When comparing military to non-military rivalries, the benefits of the latter derive from the fact that such competition is not a zero-sum game. In order for one side to win a military contest decisively, the other side must decisively lose, with all of the destruction of human lives, property, and natural resources that such a loss entails. And when neither side decisively wins, as is often the case with war, both sides are losers.

Non-military competition also involves wins and losses, but if these losses are not destructive, they cannot be regarded as part of a zero-sum game. Any individual basketball game, for example, must produce an outcome with a winner and a loser and thus appears to be a zero-sum game. But the game of basketball itself, when considered as an activity or a business, is not a zero-sum game. As an activity, basketball yields amusement and exercise to all who play it, regardless of whether they win or lose a given game. As a business, it yields financial dividends to all of the teams that participate even if it yields greater dividends to whichever team wins the most. The same principle applies to numerous forms of civilized competition, from Spelling Bees and Beauty Pageants, to competition for the Nobel Prize. As Robert Wright has argued, the *nonzero sum game* has become essential to the day to day functioning of modern civilization, even when we don't always recognize when we are playing it (2001).

Naturally, the framework for such civilized competition requires the maintenance of peace. Since 1945, at least three factors have helped to prevent open warfare among the world's major powers: first, the institutional framework for international cooperation offered by the United Nations; second, the preponderance of American power as a check on the aspirations of most military adventurists; and, third, the horror of nuclear holocaust as a deterrent to war between any states possessing those weapons. Unfortunately, each of these factors has lost some of its power to keep the peace over the course of time. The five permanent members of the United Nations Security Council (Britain, Russia, France, China, and the U.S.) offer no representation to any country in Latin America or Africa and exclude India, the world's largest democracy. The United States will likely be surpassed by China as the world's largest economy in the first half of the twenty-first century, and it will not be able to maintain its current levels of military spending indefinitely. These factors, compounded by severe social and economic challenges within American society, suggest that the preponderance of American military power far beyond its own borders will reach its expiration date well before the end of this century. The logic of nuclear weapons as a deterrent to war has also suffered a blow in the twenty-first century due to the proliferation of nuclear technology, and the advent of terrorist groups with religious aims. The proliferation of nuclear technology has made these weapons more accessible to small nations than they were in the past, thus ending the paradigm of a "cold war" between a few major powers that Orwell foresaw when he argued in 1945 that this new kind of weapon would probably lead any power which could afford to produce it into become the sort of "state which was at once *unconquerable* and in a permanent state of 'cold war' with its neighbors". As a steadily growing list of smaller and more volatile regimes have acquired nuclear weapons, however, Orwell's prediction that a nuclear stalemate among sovereign states could remain stable indefinitely has become much more tenuous. Furthermore, the proliferation of terrorist groups with religious aims has made the use of nuclear weapons more likely since many of these political actors, expecting their reward in the afterlife, will not be deterred by the logic of Mutually Assured Destruction.

In other words, the international order that has maintained relative peace between the great powers since 1945 was tenuous to begin with, and the historical circumstances that made it possible have precipitously less influence on the course of world events in the twenty-first century. It is worth noting, however, that this period did produce some considerable achievements in the area of protecting the global commons. For example, the Antarctic Treaty of 1959 was one of these achievements. It established the entire continent of Antarctica as off limits to territorial claims, and contained a strong environmental component by also forbidding the testing of nuclear weapons and the disposal of nuclear waste on that continent. Four years later, the Partial Nuclear Test Ban Treaty of 1963 ending the atmospheric testing of nuclear weapons by the United States, the Soviet Union, and the United Kingdom did little to slow the nuclear arms race, but it drastically reduced the problem of radioactive fallout in the earth's atmosphere. It also signaled a growing awareness on the part of some, including John F. Kennedy, that human activities could alter the earth in profound and unpredictable ways. As President Kennedy noted in his address to the National Academy of Sciences in April of 1963 that, "science today has the power for the first time in history to undertake experiments with premeditation which can irreversibly alter our biological and physical environment on a global scale" (Hamblin 2013, p. 147).

Of course, Kennedy's words are also applicable to those global experiments that are not premeditated, such as the impact of chlorofluorocarbons (CFCs) on Earth's protective ozone layer, or the impact of fossil fuel combustion on its climate. In contrast to the radiocative fallout from nuclear weapons tests, any attempt to lessen these impacts on a global scale will require a significant change in the behavior of not only national governments, but also private industry. The record of success in this area has been mixed at best. The Montreal Protocol that went into effect in 1989 has been largely successful in limiting the global emission of CFCs, after it was discovered and confirmed by satellite imagery that these chemicals, used as refrigerants and aerosol propellants, were eroding the ozone layer. The Montreal Protocol to protect the ozone layer was one of the few international agreements that effectively regulated the activities of private industry on a global scale, but its success was probably due to a unique set of circumstances. The creation of an international protocol curtailing the production and use of CFCs was a remarkable success in the history of environmental regulation, but it coincided with trends and events in the chemical industry that eased its path toward implementation. The DuPont Company, as the world's dominant producer of CFCs, could have presented formidable opposition to the ratification of the Montreal Protocol in the United States, but ultimately chose not to because the implementation of a new regulations regime coincided with its own business interests. Even before the Montreal Protocol was negotiated, the production of the chemicals CFC 11 and 12 (commonly known by the DuPont trade name Freon) were posing challenges to DuPont's business strategy. As MIT researchers James Maxwell and Forrest Briscoe observed: "By the mid 1980s, production of CFCs 11 and 12 was no longer as profitable a business as it once was." In a market that was "already characterized by overcapacity," a dominant producer such as DuPont could enhance its environmental image while pursuing policies that were in fact quite helpful to its bottom line: "International regulations offered major producers the possibility of new and more profitable markets in the long term" (Maxwell and Briscoe 1997).

The success of the Montreal Protocol in limiting damage to the ozone layer from CFCs owes something to both a willingness on the part of governments to recognize a serious threat to the global commons, and to the fortuitous fact that the largest single producer of CFCs saw an opportunity, in this case at least, to advance its own business interests by marketing alternatives to those chemicals that had been identified as an environmental hazard. It would be unwise, however, to expect that multinational corporations will always find such handy opportunities to do well while doing good. For example, the damage that the mining and combustion of coal does to the land, oceans, and atmosphere is staggering. Coal extraction leads to the destruction of mountains, the poisoning of rivers and water tables, and the combustion of coal not only contributes to the greenhouse effect but helps to drive the process of ocean acidification. However, commercial interests involved in mining coal are not going to find it profitable to shift to the production of an alternative source of energy. The equipment and facilities that they have invested in the extraction, processing, and combustion of coal represent an investment that cannot easily be repurposed. In contrast to DuPont's willingness to find substitutes to CFCs, those businesses heavily invested in coal, such as Koch Industries, have become major backers of efforts to deny scientific evidence concerning the dangers to public health and the environment posed by coal. Determined to characterize over a century of research on the connection between carbon emissions and climate change as "junk science" and to paralyze efforts to regulate greenhouse gas emissions, Koch Industries has used its considerable resources to influence Congress: "Koch spent $12.3 million on lobbyists in 2009, ranking it fifth behind Exxon, Chevron Corp., ConocoPhillips Co. and BP PLC" (Lehmann and Climatewire 2010). If the U.S. Congress is weak enough to be manipulated by a coterie of fossil fuel corporations, then we must build a more powerful legislative body that can compel these entities to recognize basic science and submit to the rule of law. If we are to protect our oceans, atmosphere, and climate, the power of the fossil fuel industry must be met with the creation of new democratic institutions that will be more effective than any national legislature, because they will operate, as these corporations do, on a global scale. As Elisabeth Mann Borgese foresaw, the only way to transcend the Tragedy of the Commons, and to protect "the common heritage of humanity" is to establish democratic accountability and the rule of law beyond the borders of the nation state.

## Bibliography

Baratta, Joseph Preston. 2004a. *The politics of world federation*. Vol. 1. Westport: Praeger.

———. 2004b. *The politics of world federation*. Vol. 2. Westport: Praeger.

Bender, Marilyn. 1969. The lone woman in think tank is a 'Renaissance Man.' *New York Times*. May 27, 1969. 50.

Borgese, Elisabeth Mann. 1965. *A constitution for the world*. Santa Barbara: Center for the Study of Democratic Institutions.

———. 1969. "*Lecture on the ocean regime*". File MS-2-744, Box 139, Folder 16 Elisabeth Mann Borgese archives. Dalhousie University.

———. 1999. Global civil society: Lessons from ocean governance. *Futures* 31 (1999): 983–991.

Boyer, John W. 1995. "*Drafting salvation*" University of Chicago Magazine. Vol. 88, No. 2.

Cronon, William. 1996. *Uncommon ground: Rethinking the human place in nature*. New York: W. W. Norton & Co.

Diamond, Jared. 1999. *Guns germs and steel: The fates of human societies*. New York: W.W. Norton & Company.

Hamblin, Jacob Darwin. 2013. *Arming mother nature: The birth of catastrophic environmentalism*. Oxford: Oxford University Press.

Hesiod. 2017. *The poems of hesiod: Theogony, works and days, and the shield of herakles* trans by Barry B. Powell. Oakland: University of California Press.

Holden, Barry. 2002. *Democracy and global warming*. London: Continuum.

Jakhu, R.S., J.N. Pelton, and Y.O.M. Nyampong. 2017. *Space mining and its regulation*. Cham: Springer.

Lehmann, Evan, and Climatewire. 2010. Who funds contrariness on climate change? In *Scientific American* http://www.scientificamerican.com/article/who-funds-contrariness-on/.

MacAskill, Ewen. 2008. "Hurricane gustav: Republican convention thrown into chaos" *The Guardian*, 31 August, 2008. https://www.theguardian.com/world/2008/sep/01/usa.republicans2008

Mackaye, Benton. 1951. "Toward Global Law". First published in *The Survey*, Vol. LXXXVII, No. 6. June, 1951) Republished in *From Geography to Geotechnics* (1968). Champaign: University of Illinois Press.

Mann, Thomas. 1901. (1983). *Buddenbrooks*. trans. H. T. Lowe-Porter New York: Alfred A. Knopf.

Maxwell, James, and Forrest Briscoe. 1997. There's money in the air: The CFC ban and Dupont's regulatory strategy. *Business strategy and the environment* 6.

Mazower, Mark. 2012. *Governing the world: The history of an idea*. New York: The Penguin Press.

Monnet, Jean, Richard Mayne, trans. 1978. Memoirs. New York: Doubleday & Company.

Ricks, Thomas. 2017. *Churchill and Orwell: The fight for freedom*. New York: Penguin Press.

Safire, William. 1982. "Come to 'Club Seabed'". *The New York Times*. November 8th, 1982.

Ward, Barbara, and René Dubos. 1972. *Only one earth*. Harmondsworth: Penguin.

Weber, Max. 1919. (2004). *The vocation lectures*. Indianapolis: Hackett Classics.

Wright, Robert. 2001. *Nonzero: The logic of human destiny*. New York: Vintage Books.

# Chapter 6
# Governing Ourselves

*"Democracy needs support and the best support for democracy comes from other democracies. Democratic nations should... come together in an association designed to help each other and promote what is a universal value — democracy."*

Benazir Bhutto

**Abstract** The rapid integration of commerce, transport, and communication over the past century seems to suggest that the world is on the path to becoming a single global civilization. Although the form of governance that will maintain that civilization has yet to be determined, there is no reason to assume that it will be democratic, unless there is a deliberate and sustained effort to assure that it is. Given the symbiotic relationship between science and democracy, and the superior historical record of democratic governments on environmental issues, there is reason to believe that the establishment of democratic governance on a global scale would offer the best strategy for dealing with climate change.

**Keywords** Timothy Ferris · Human rights · Judith Shapiro · Symbiotic civilization

The second half of the twentieth century cleared a path for the establishment of democracy, the rule of law, and environmental protection on a scale unprecedented in all prior history. The success of agreements such as the Atmospheric Test Ban Treaty and the Montreal Protocol owed much to the international framework for dialogue and negotiation established after 1945, but that framework has weakened considerably in recent decades. These global accords were not strictly products of the environmentalist movement, but emerged from a complex nexus of concerns involving such diverse priorities as global security, human rights, and public health. These accords illustrate what is possible when environmental issues are not treated separately from other human needs and aspirations, and they also suggest that more effective global accords addressing climate change are not impossible.



R. S. Deese, *Climate Change and the Future of Democracy*, Environmental Challenges and Solutions 5, https://doi.org/10.1007/978-3-319-98307-3_6

Unfortunately, by the time Kyoto Protocol was established in 1997, the international framework that had made it possible was showing new signs of weakness. The United States, which had laid the foundations for so many international institutions after World War Two, was undergoing political changes, especially in Congress, that made it much less likely to commit to the Kyoto Protocol. The challenges of the Second World War and the Cold War had compelled the United States to abandon isolationism, and, for nearly half a century following Pearl Harbor, the United States helped to build and maintain a number of important international institutions and alliances, including the United Nations (UN), the North Atlantic Treaty Organization (NATO), and the Organization for Economic Cooperation and Development (OECD). In the wake of the Cold War, however, the tenor of American politics once again retreated from the spirit of internationalism. Because the United States was now so obviously dependent on natural resources, investment, and manufactured goods from overseas, the old American habit of isolationism was no longer tenable. Unilateralism took its place. According to this school of thought, the United States would not renounce involvement in overseas affairs and conflicts as the isolationists had attempted to do in the 1930s. Unilateralism dictated that the United States would continue to intervene militarily around the world, but that it would regularly exercise the right to do so entirely or primarily on its own, acting outside the framework of global institutions such as the United Nations Security Council, or even regional alliances such as NATO. The single greatest example of this approach to world affairs was the 2003 U.S. invasion of Iraq, but this major military campaign was preceded by the 1989 U.S. invasion of Panama, and a mounting number of unilateralist U.S. airstrikes and bombing campaigns in the 1980s and 1990s against Libya, Sudan, Iraq, and Afghanistan. The spirit of unilateralism not only affected the military policy of Presidents from both political parties, but also changed the approach of the United States Congress to the ratification of international treaties, including those related to the global commons. Reflecting the durability of what the historian Richard Hofstadter famously dubbed “the paranoid style in American politics” in 1964, many international institutions such as the United Nations continue to be regarded as “un-American” by a vocal minority of U.S. citizens, even though the United States had been instrumental in building them (Bess 2006. 283). During the Cold War, this tendency was mitigated by the common cause of anticommunism that united both Democrats and Republicans and served as a rallying point for both internationalists and unilateralists. With the disappearance of the Soviet threat, unilateralist voices become more numerous and increasingly shrill in American politics, descrying even such agreements as the United Nations Convention on the Law of the Sea (UNCLOS) as a sinister attempt to impose an Orwellian “new world order” on the people of the United States. In this climate of suspicion regarding international treaties, the Senate has failed to ratify UNCLOS, even though presidents from both parties have called on it to do so since 1994.

In light of this American tendency toward unilateralism, the fate of the Kyoto Protocol in the United States should not have been surprising. Though signed by President Clinton in 1997, it faced considerable opposition in the Senate for the remainder of his presidency, delaying its submission for ratification by the necessary two-thirds majority. When Clinton's successor, George W. Bush, came into office, he immediately withdrew the United States from the Kyoto Protocol entirely, thus setting the stage for other nations such as Canada and Russia to follow suit. While it should always be conceded that reasonable people may not agree about a given treaty such as UNCLOS or the Kyoto Protocol, it must also be noted that much of the opposition to both of these international treaties has been fueled by conspiracy theories, pseudoscience, and a particularly unhinged brand of xenophobia. From this combination of factors, opponents of multilateralism have gone beyond arguing the idea that the ratification of the Kyoto Protocol would be bad for the United States, and proliferated claims that climate change itself is an elaborate hoax designed to serve the nefarious interests of a cabal of scientists and internationalists.

Senator James Inhofe of Oklahoma has even written a book on climate change to promote this view, entitled *The Greatest Hoax: How the Global Warming Conspiracy Threatens Your Future*. Inhofe's ties to the fossil fuel industry are extensive. According to the non-partisan Center for Responsive Politics, Inhofe has received over 1.8 million dollars in direct support from fossil fuel producers in the two decades between 1998 and 2018, with Koch Industries and Murray Energy ranking as his largest contributors (Open Secrets 2018). The systematic denial of the link between industrial pollution and climate change in some ways resembles the earlier campaign of denial by tobacco companies in the face of mounting evidence about the health dangers posed by their product. As the historians Naomi Oreskes and Erik M. Conway have documented, both industries have employed the same tactics, and sometimes the same public relations firms, to execute their strategies of denial, obstruction, and delay (Oreskes and Conway 2011). Because Inhofe and many others who deny the danger of climate change paint international attempts to reduce greenhouse gas emissions as part of a global conspiracy against the United States, their rhetoric relies upon and reinforces "know-nothing" and nativist themes that have long been a part of American politics, and which have found a growing audience since the end of the Cold War.

In addition to the growing influence of unilateralist politics in the United States, a major shift in the global balance of power has also made international cooperation on environmental issues more difficult. The single greatest factor in this shift has been the growing economic and political clout of China since the 1980s. In the heyday of the Cold War, agreements such as the Partial Nuclear Test Ban Treaty represented a compromise between the United States and its allies on one side, and the Soviet Union and its allies on the other. By the 1990s, the negotiations for agreements such as the Kyoto Protocol reflected a quite different division of power and interests, with the People's Republic of China and other newly industrialized economies representing one side of the table, and the United States and other mature industrial economies on the other. On the surface, the argument presented by China and other industrializing countries was that they should not be held to the same goals in reducing greenhouse gas emissions because they had not been polluting the atmosphere for nearly as long as countries that had industrialized more than a century prior to China. Ostensibly, this was the reason that newly industrialized countries were held to a much less ambitious standard for greenhouse gas emissions reduction in the Kyoto Protocol. However, since so much industrial production for U.S., European, and Japanese companies takes place in newly industrialized countries such as China, it could also be argued that the mature industrial countries were also lowering the greenhouse gas emissions targets for their own major corporations as they steadily moved their operations to China. In American politics, the exemption contained in the Kyoto Protocol for newly industrialized nations was presented as unfair to the United States, while in actual practice that exemption assured that the outsourcing of U.S. industrial production to China would continue to be a bargain, not only as a result of cheap labor costs, but also as a result of lax or nonexistent environmental regulations. In other countries with mature industrial economies, such as Germany and Japan, allowing China to adhere to a lower standard for the reduction of GHG emissions presented the same opportunity to raise their standards at home while shifting a growing share of their manufacturing to China, where they could operate outside of those standards.

Although the scientist and climate activist James Hansen has praised the government of mainland China for its acknowledgment of climate change and its commitment to renewable energy, he also observes China is heavily dependent on coal for its "baseload electric power" and argues that it will have to "phase out coal emissions" if its commitment to addressing climate change is to have any measurable impact (Hansen 2009, p. 192). In fact, Hansen makes the case that if "coal emissions are phased out rapidly" across the world, "the climate problem is solvable" (Hansen 2009, p. 173). In summarizing his vision for how to address the climate crisis, Hansen calls for a global energy regime that would "disallow the use of coal and unconventional fossil fuels (such as tar sands and oil shale) unless the resulting carbon is captured and stored" (Hansen 2009, p. 205). However, it is impossible to enforce this level of regulation through agreements as weak as the Kyoto Protocol or the Paris Climate Accords. In spite of its commitment to expanding the use of renewables, China has rapidly increased its used of coal since 2009. By 2014, "China burned as much coal as the rest of the world combined" (Koehn 2015, p. 23). Fragile treaty arrangements such as Kyoto or Paris cannot slow this juggernaut of fossil fuel consumption, by China, India and other rising industrial economies. What is required is the establishment of a global framework of law that transcends national borders. And, in order to be legitimate and enduring, that framework must be democratic in its origins.

In light of the increasing mobility of capital since the 1980s, the nation-state paradigm for international agreements affecting the global commons has become far less applicable to the facts on the ground. In 1959, the sovereign governments with an interest in the fate of Antarctica were able, in spite of their ideological differences, to reach a treaty that was fairly clear-cut because governments remained the primary sponsors of the exploration and scientific research being conducted on Antarctica. In 1963, it was also possible for the Eastern and Western blocs to come to an agreement on the Partial Nuclear Test Ban Treaty because the respective governments on each side tightly controlled the production and testing of nuclear weapons. When it comes to drafting an enforceable agreement to lower carbon emissions, national governments may have less real influence than multinational corporations. By virtue of the increasing mobility of capital, these private interests are never strictly accountable to any single government. Since all governments seek the revenue generated by multinational corporations, no single government is willing to implement regulations that would drive all or most their business elsewhere. This phenomenon, which has come to be known as the "race to the bottom," has thwarted the creation of standards to regulate the environmental impact of multinational corporations. The creation of a democratic institution larger than any single nation state offers the only effective way to curtail the activities of multinational corporations and address the challenge of climate change.

When the historian Benedict Anderson coined the term "imagined communities" to describe modern nation states, he pointed to a truth that is at once universally obvious and universally ignored (1983). The question of whether human civilization can survive the challenges posed by climate change is inextricably entangled with the question of whether we can come to recognize the fictitious nature of the nation state. The mosaic of borders that divides the people of the earth into sovereign nation states is of course invisible from space, but it would be facile to claim that its invisibility made it less than real in terms of its effect on human behavior. Much of the human behavior that is visible from space, however, does not seem much affected by national borders. Civil aviation is an excellent example of this. The first planes with pressurized cabins and jet engines were created to wage war between empires and nation states, but the civil aircraft descended from these warplanes represent the real quantum leap in the history of humanity. By revolutionizing personal transport and global commerce, they have steadily made empires and nation states far less relevant than they once were. The civil aviation network that has grown to encircle the earth heralds the birth of a single system of governance.

A look at the daily transit of people and goods through the sky, across the oceans, and across the continents of the earth would suggest a number of things to an observer from space. First, this observer would conclude that the species responsible for all of this transit had mastered some form of energy, not only to move so many people and things but to speckle the continents and islands of this planet with galaxies of glittering lights that spring to life as each side of this rotating globe turns away from the light of the sun and into darkness. If they were informed that most of this light comes from fire, i.e. the burning of fossil fuels, our alien observers might be somewhat less impressed with our technological prowess and ecological common sense at this point in our development, but they would nonetheless be likely to conclude that our globe had been enveloped by a vast system in which collaborative effort had become the norm. The structural patterns of our cities and farms, as well as the daily transit of planes, ships, trains, and automobiles across the globe, might appear to be something like a network of beehives or termite colonies to an observer from space. Countless systems are operating in every habitable part of the globe, and they seem to be coalescing into a single system. From the ground, we are aware of starkly different religions, philosophies, and systems of government that operate around the world, but from space the patterns formed by our cities and suburbs look increasingly the same and increasingly enmeshed in a single network. While traumatic events on the ground have caused human beings in various parts of the world to worry about what Samuel Huntington called "a clash of civilizations" the view from space seems to suggest that the human race is accelerating its trajectory towards a single, all-encompassing global civilization (2011).

The chess champion and human rights activist Garry Kasparov describes the threats to peace and democracy that have emerged since 1989 as forms of reaction against the emergence of a global civilization that honors individual rights. As Kasparov sees it, jihadist terrorist groups such as ISIS and reactionary autocrats such as Vladimir Putin share a fundamental trait: they are "time travelers" struggling against "the twenty-first century's borderless ideas and technologies" as they seek to "hold onto their waning authority by stopping the advance of the ideas of an open society" (Kasparov 2015, p. 254). Combatting such reaction will provide a sustained commitment to democracy, individual rights, and freedom of inquiry. Kasparov concludes that the defense of these values against their violent and dictatorial opponents is a struggle that transcends geography, culture, and nationality: "What unites the time travelers is their rejection of modernity, their fear and hatred of what we should simply call 'modern values' to replace the obsolete and condescending term 'Western values'" (Kasparov 2015, pp. 254–255). Since a narrow victory in the Electoral College put Donald J. Trump in the White House, Kasparov has warned of his similarities to, as well as his affinity for, Vladimir Putin. Six months into Trump's presidency, the longtime dissident and advocate for democracy warned that Trump was borrowing more than a little from the Putin playbook: "There is a clear parallel here to what we have experienced in Russia for the past 17 years under Putin: the intentional conflation of the private interests of the few with the public good" (Kasparov 2017).

At various times in history, people have perceived the nascent growth of an interdependent global civilization, and some have celebrated it. In his brilliant book on the expansion of telegraph networks in the nineteenth century, Tom Standage gathered together the proclamations of many who celebrated such milestones as the first transatlantic telegraph cable in 1858 as evidence that a new millennium of peace and prosperity for all was just around the corner (1998). In the twentieth century, the paleontologist and Jesuit visionary Teilhard de Chardin had a similarly poetic vision of what he called the "noösphere" and many of his ideas were further developed by Marshall McLuhan, who coined the term "the global village" to describe the convergence of politics and popular culture on a global scale that electronic media had made possible. The prospect that humankind is probably on the road to forming a global civilization should not lead us to assume, however, that this emergent civilization will necessarily be prosperous, happy, and free. Without the establishment of democratic accountability on a global scale, there is every possibility that this emerging civilization could be a system of global exploitation that creates poverty, misery, and new forms of slavery for the many while it accumulates wealth for the few.

The terms "symbiosis" and "civilization" may belong, respectively, to the fields of biology and history, but their roots are inextricable. One of the most commonplace examples of symbiosis is the lichen, which combines the photosynthesizing power of algae and/or cyanobacteria with the fibrous structure of fungus. In this case, one species offers the production of chemical energy from sunlight, while other provides some shelter from the surrounding elements. A well-known example of symbiosis in the animal kingdom is the relationship between the African Oxpecker Bird and the African Rhinoceros. In this case, the bird eats ticks and other insects it finds on the back of the Rhino, thus ensuring a steady diet for itself and eliminating a host of parasites from the larger animal. Many biologists have argued that this relationship offers more benefit to the bird than to the Rhino, but it still results in a net benefit for both species. The earth's biosphere contains a bewildering variety of symbiotic relationships. Some of these, such as the relationship between the Oxpecker and the Rhino, are contingent enough that each animal could survive without the other. Other symbiotic relationships are obligating, as in the case of many varieties of lichens in which neither of the component species would be able to survive on its own.

Before it was applied to plants and animals, the word symbiosis had long been used to describe human communities. The Oxford English Dictionary defines symbiosis as a synonym for companionship and cites the first use of the word in 1622 by the merchant Edward Misselden. The goal of his book *Free Trade*, declared Misselden, was "To study and inuent [*sic*] things profitable for the Publique [*sic*] Symbiosis." Since the nineteenth century, however, the term has been more commonly used to describe strictly biological rather than social phenomena. While symbiosis is still used to described examples of interdependency between various plants and animal species, other terms are used in the realm of human affairs, such as politics, economics, and international relations. Each of these terms has a specificity that is certainly useful, but all are subsets of the broader concept of symbiosis. When we contend here and concur there concerning a political, economic, or international question, we are debating how we can continue to live together. In light of this simple fact, Clausewitz's famous dictum that "War is politics by other means" is a bald falsehood. Politics, however ugly it may be, is the art that human beings have contrived to live together in complex societies, and is therefore a form of symbiosis. War abandons symbiosis in favor of lethal force and can only end with the death of all parties concerned, or the resumption of symbiosis. When all parties in a war are ready to set killing aside and return to the pursuit of symbiosis, the practice of politics will resume again.

In his *Politics*, Aristotle observes that "the human being by nature is a political animal" and cannot survive outside of a complex web of relationships with other human beings (Riesbeck 2016, p. 2). If we consider that the advent of planting crops and raising livestock made the first cities possible, it then becomes apparent that human civilization may represent a form of symbiosis in the very broadest sense of that term. Beyond being dependent upon each other through political and economic relationships, human beings have built complex and interdependent relationships with other species, and thus created an obligatory form of symbiosis in which a number of crop and livestock species depend upon us for survival at the same time that the survival of our civilization depends on them. Human civilizations therefore furnish examples of both social and biological symbiosis. Or, to put it in simpler terms: human beings must live in communities with other human beings in order to survive, and those communities must cultivate ongoing relationships with other species of plants and animals in order to ensure their survival. There are many truisms that we ignore at our own peril, and this is one of them.

The challenge of climate change will compel humanity to experiment with new models of government on a global scale. As we begin this process of experimentation, we should consider the evidence that democracy is a more ecologically sustainable form of government than authoritarian rule. In light of that evidence, we should also explore how it might be possible for institutions of global governance to be more democratic. Nation states that are held accountable by the rule of law, a free press, freedom of conscience, and a system in which multiple parties can compete in free and fair elections for a share of political power have a better environmental record than nation states that do not face this kind of accountability. On a broader scale, international federations that include electoral representation, such as the European Union, have a better environmental record than trade confederations that do not include electoral representation, such as the North American Free Trade Agreement. The first of these principles is well illustrated by the contrast between the environmental record of the Soviet Union and the United States in the twentieth century, in which the latter polity pursued much sounder environmental policies than the former. The second of these principles is well illustrated by the environmental record of the European Union and the members of the North American Free Trade Agreement in the twenty-first century, in which the democratic federalism of the EU has produced a much more sustainable policy toward both greenhouse gas emissions and agriculture than has been achieved in North America (Carlarne 2010, p. 241). This suggests that democratic federalism on a supranational scale is a boon to ecological sustainability. Of course, if we accept the premise that democracy is green, we must face the question of its mortality. Like other green things, it is also fragile. Every place on earth where democratic governance exists today still faces the danger that it might disappear tomorrow. Every place on earth where democratic governance is deferred presents a challenge to democracy in every other place on earth. As the founder of the Green Belt movement Wangari Maathai declared in her 2004 Nobel address, "Recognizing that sustainable development, democracy, and peace are indivisible is an idea whose time has come" (Tal, ed. 2006, p. 255).

In the late eighteenth century, when Immanuel Kant argued that democratic federalism was the best system for preserving the peace between nations, the dangers of environmental calamities such as climate change had not yet been imagined. Now that we face those dangers, it is worth questioning whether Kant's vision of democratic federalism might prove useful not only for preserving peace between nations, but also peace between our own species and the rest of the biosphere. Kant's claim that democracies are far less likely to pursue wars with other democracies has been called Democratic Peace Theory. The historian Spencer Weart conducted a thorough and systematic investigation of Democratic Peace Theory and concluded that it is quite well supported by the record of history. Assessing the probability that "the absence of war between established democracies is a mere accident" Weart reached that conclusion that the odds against this were "less than one chance in a thousand" (Weart 1998, p. 4). Because of the very high likelihood that there is a causal relationship between democracy and peace, Weart argued for a stronger commitment to the spread of democracy: "If democracy is to survive and expand, many people must devote their lifetimes; many must give their very lives, as millions before have done." Weart acknowledges that success of efforts to establish "universal democracy" was far from certain, but reasoned that, "the goal is worth even more than has commonly been supposed. For if we can attain this, we will at the same time attain universal peace" (Weart 1998, p. 296).

Although no one has investigated the question as thoroughly as Weart has studied Democratic Peace Theory, it is also possible that democratic societies are more likely to be at peace with nonhuman nature than their authoritarian counterparts. If we expand upon that premise and suppose that democratic states are also less likely to ravage the earth's biosphere than nondemocratic states, we might call such a premise Democratic Sustainability Theory. As with any theory about political arrangements, this is not a claim that can be proven once and for all, though the ability of democratic societies to identify problems and correct them over time has tended to make them more responsive to environmental problems. The scholar David Runciman, though acutely aware of the shortcomings that have bedeviled democracies over the past century, points to their proven record of success on environmental issues, declaring, "Autocratic regimes are the worst polluters and the greatest squanderers of natural resources," while "democracies make far better use of their resources because they are far more resourceful" (Runciman 2013, p. 315). It is important to emphasize, of course, that all claims made here for the sustainability of democracies are relative and not absolute. To paraphrase Churchill, we might look at the environmental misdeeds of democratic societies and conclude that Democracy is the least sustainable system of government – except for all the others.

Of course, even a relative claim for democracy as the "least bad" option for sustainability requires a serious defense. Why should we expect a greater degree of democratic accountability to improve the sustainability of human civilization, or at least make it less unsustainable than it is now? The most fundamental reason why democratic governance is likely to be more ecologically sustainable than authoritarian governance has been well articulated by the China scholar Judith Shapiro. In her survey of environmental degradation during the Maoist era, Shapiro states that:

> While it is far from clear that only democracies can create sustainable relations with nature, pluralism permits the expression of ideas in open public settings and in print, promoting criticism and debate. It allows people the opportunity to disagree, to seek and offer additional information, to deliberate, and to modify and improve proposals that have impact on the natural world. With such freedoms, moreover, there is a better chance that the unintended consequences of human activity will be revealed in a timely fashion. Basic human freedoms thus deeply affect human actions vis-à-vis the nonhuman world. (Shapiro 2001, p. 65)

By arguing that democratic pluralism is likely to improve our understanding of our relationship with the "nonhuman world," Shapiro sheds light on the same fundamental connection between science and democracy that Chinese intellectuals such as Chen Duxiu had proclaimed in his seminal essay on "Mr. Democracy and

Mr. Science" in January of 1919 (Calhoun 1997, p. 317). Science and democracy cannot be predicated on arbitrary authority, and must by definition allow the flow of new data and analysis from all quarters. Thus, a person of humble origins can rise to a position of leadership in a democracy, or overthrow a dominant paradigm in a field of science – as illustrated, respectively, by the careers of Lincoln, who started out as a rail splitter from Kentucky, or Einstein, who published his most revolutionary theories while working as a patent clerk in Zurich.

While science and democracy share fundamental similarities and goals, it must be conceded that there have been many cases where democratic societies have been slow to recognize and respond to important scientific discoveries. While this has certainly been the case with climate change, the denial of scientific evidence in the service of short-term economic interests is far from new. Henrik Ibsen's seminal play *An Enemy of the People* offers a brilliant study of what can happen when a scientifically verified truth is diametrically opposed to the short term economic interests of a democratic community. Ibsen's protagonist is a young doctor named Thomas Stockman who lives in a small Norwegian town that derives its wealth from its local hot-springs and baths, famed for their curative powers. When Thomas investigates the purity of the water in the town baths, he discovers that a recent re-routing of the pipes under a local tannery has led to a serious contamination of the baths. As a successful doctor and the younger brother of the mayor, Thomas sees himself as both an advocate of progress and a pillar of his community. Convinced that "the free press" and "the solid majority" will support his efforts to protect public health, Thomas expects to be treated like a hero for bringing this important discovery to light (Ibsen 1882 2007, p. 45). To his dismay, he soon finds that virtually every person of influence and authority in the town, especially his older brother, is imploring him to suppress his discovery and threatening to ostracize and defame him if he does not. By the close of the play, Thomas and his family are thoroughly alienated from all their friends, neighbors, and relations, and the young doctor who had imagined that he would be hailed as a hero for his efforts to protect public health has been branded "an enemy of the people." Written in the late nineteenth century, *An Enemy of the People* reflects a time when the role of bacteria as a disease vector, though well understood by scientists, had still not been universally accepted. In the United States for example, President Garfield died of a gunshot wound that would not have otherwise been fatal simply because the doctors treating him remained willfully ignorant concerning the germ theory of disease and had not adopted the practice of sterilizing their instruments (Mirsky 2017). Throughout *An Enemy of the People* Thomas Stockman encounters a similar incredulity about the invisible creatures that he has found in the water of the town baths.

Within the relatively narrow parameters of time and space in which Ibsen sets his story, the interaction between science and democracy ends in disaster. Ideally, Thomas Stockman's scientific discovery that the town baths have been contaminated with dangerous bacteria should be a boon to the community. After all, the sooner they can re-route the pipes to the baths and restore the purity of their waters, the sooner they can save lives and protect the reputation of their town. For this reason, Thomas at first places an almost childlike faith in the democratic institutions of his community, such as its newspaper, its town-hall meeting, and its mayor. And yet, by the last act of the play, the truth that he has exposed through careful science has been ignominiously rejected by the local institutions of democracy. As he grows angry at the townspeople and especially his own brother, Thomas speculates that all of them are ill bred and that society is rife with dimwitted mongrels. His offhand suggestion that society would be better off if people were bred scientifically also mirrors the time in which the play was written, as the concept of eugenics was then garnering a growing influence in many Western societies. Early in the play, Thomas mischievously grabs his brother's top hat and declares "I am king of the town now" (Ibsen 1882 2007, p. 82). Though they may seem at first like small details, the young doctor's musings on the deliberate breeding of an improved citizenry or his juvenile joke about seizing his brother's position of power are pregnant with historical meaning.

The concept of eugenics, though it would suffer a serious loss of prestige after World War Two because of its infamous connection to the Third Reich, had been widely accepted by intellectuals in the early twentieth century. In fact, many early and influential figures in the environmental movement, such as Julian Huxley in Britain and Fairfield Osborn in the United States, were lifelong advocates of eugenics. As noted earlier here, Garret Hardin, whose essay on "The Tragedy of the Commons" is one of the most anthologized and cited texts in the history of environmentalism, was an unapologetic advocate of eugenics and extensive state intervention in human reproduction. Just as Thomas Stockman became disillusioned with the "damned majority" of the town that he was trying to protect, many environmentalists in the twentieth century became disillusioned with the indifferent or hostile response of the press and public to their warnings about the dangers of nuclear technology, overpopulation, or climate change (Ibsen [1882] 2007, p. 100). Throughout the twentieth century, some environmentalists have dreamed of a society in which scientifically informed individuals such as themselves have seized the mantle of power, just as Thomas playfully seized his brother's top hat. And they have dreamed, just as Thomas did, of breeding a public which they imagined would be more intelligent and therefore more receptive to scientific truth.

In *The Science of Liberty*, Timothy Ferris offers a perceptive and devastating critique of both eugenics and the notion that society would be better run if a cadre of scientists were in charge. Ferris points out that Francis Galton, the man who coined the term eugenics, undermined his own theory with his empirical observation of a group of bettors at a country fair. When a large group of fair-goers were invited, for a small betting fee, to record their wagers on the weight of a prize pig, Galton predicted that the average of their estimates would be extremely close to the truth. This early hypothesis about what has come to be called "the wisdom of crowds" has proven to be correct, and the accuracy of such aggregate predictions has proven to become greater in direct proportion to not only the size, but also the diversity, of the group making estimates. The attendees that Galton saw at the fair that day had not been bred by an elite caste of scientific experts, and their aggregate ability in practical matters illustrated they did not need to be ruled by such a guardian caste of technocrats (Ferris 2011, pp. 31–31).

*An Enemy of the People* remains a brilliant parable for the complex relationship between science and democracy. Ibsen's depiction of a popular mayor seeking to suppress an important discovery in order to safeguard local business interests would become a major inspiration for Peter Benchley's 1974 novel *Jaws*. Ibsen also inspired Benchley's screenplay for the subsequent blockbuster movie, which director Steven Spielberg described as, "*Moby Dick* meets *An Enemy of the People*" (Krasilovsky 2017, p. 66). In Benchley's tale, the protagonist is a police chief who realizes that the New England beach town of Amity where he now works has become the hunting ground for an especially large great white shark. The mayor of the town, like the mayor in Ibsen's play, wants to shut the police chief up in order not to imperil the summer tourist season on which the entire town's economy depends. In this case, however, it is science that resolves the impasse. When a marine biologist from the Woods Hole Institute studies the bite marks on one of the shark's recent victims, he confirms that she was killed by a great white of considerable size. The mayor can no longer deny that the town of Amity has a shark problem on its hands, and the citizens at the town hall resolve to do something about it.

When science and democracy operate in a wider field, they become more symbiotic. There is a very simple reason why important scientific information is successfully suppressed by the mayor and townspeople in Ibsen's *An Enemy of the People* and why it is ultimately accepted in Benchley's tale about the shark and the beach town of Amity. In Ibsen's play, the entire field of action is the town. It is quite likely that if an official from the provincial or national government were to learn about the impurities in the local baths, the information would be acted upon. In Benchley's *Jaws*, the individual who confirms the presence of a great white in the local waters is a marine biologist from outside the town. Not only has he scientifically confirmed that the town has a serious problem that it needs to deal with, but his mere presence confirms that this problem is no longer a secret from the outside world. The truth is not provincial, and in Benchley's story, it cannot be suppressed for long by provincial authorities.

The Chinese astrophysicist Fang Lizhi saw a strong parallel between the universal validity of scientific truth and the universal validity of democracy and human rights. In 1989, after the Chinese Communist Party had violently suppressed the democracy protests in Tiananmen Square, the government routinely advanced the argument that China had to protect its own political traditions, and its own conception of human rights. In response to this argument, Fang saw a parallel between the resistance of the CCP to democracy and the earlier resistance by imperial officials to modern astronomy:

> Recent propaganda to the effect that "China has its own standard of human rights" bears an uncanny resemblance to pronouncements made by our eighteenth-century rulers when they declared that "China has its own astronomy." The feudal aristocracy of two hundred years ago opposed the notion of an astronomy based on science, and refused to acknowledge the universal applicability of modern astronomy, even that it might be of some use in formulating the Chinese calendar. The reason they opposed modern astronomy was that its laws, which pertain everywhere, made it apparent that the "divine right of rule" claimed by these people was a fiction. By the same token, the principles of human rights, which are also universal, make it clear that the "right to rule" claimed by their successors today is just as

> baseless. This is why in every era rulers buoyed up by special privilege have opposed the idea of equality that is inherent in such universal ideas. (Fang 1989)

Pointing to the absurdity of the claim that astronomy differed across national or imperial borders, Fang argued that it was equally absurd to claim that people's fundamental rights and their aspirations for democracy were valid in one part of the world and not in another. Just as we all live under the same sky, we all have the same right to a government that derives its legitimacy from the consent of the governed. For Fang, basic literacy in science and respect for democracy went hand in hand.

If we are to consider which type of government may do the most to foster scientific literacy in the industrial age, we should recognize democracy as the form of government that has done the most to promote the twin values of transparency and accountability. If transparency and accountability are essential to the practice of scientific research, they are also essential to the sane use of applied science by any society. A number of historical examples suggest a strong correlation between political transparency and accountability and the more sustainable use of science and technology. An especially strong example of this correlation can be seen in comparing the use of nuclear technology by the United States and the Soviet Union after the Second World War. Both governments caused considerable damage to the earth's environment and to human health in their drive to exploit nuclear energy for both military and civilian applications during the Cold War, but the record of the Soviet Union is far worse. While the United States carried out numerous atmospheric tests of nuclear weapons in the forties and fifties, the Soviet Union conducted some of the most destructive tests of the 1960s, including the infamous "Tsar Bomb" test in 1961 and the Chagan test (in open violation of the 1963 Atmospheric Test Ban Treaty) in 1965. And while the record of civilian nuclear power in the United States was marred by the Three Mile Island accident of 1979, the Soviet nuclear power program produced the disastrous Chernobyl meltdown of 1986. While the Three Mile Island accident produced no casualties, the Chernobyl meltdown directly caused around fifty deaths and released radiation that will lead to the early death of around 4000 according to the World Health Organization (WHO 2005).

In the case of both military and civilian uses of nuclear technology, the social and political mechanisms of transparency and accountability have been essential to protecting the environment and saving lives. In the United States, politically active scientists such as Barry Commoner and Linus Pauling were both innovative and indefatigable in their efforts to alert the public to the dangers of nuclear bomb tests. Commoner collected thousands of baby teeth in the 1950s and proved conclusively that the radioactive isotope Strontium 90 was finding its way into the bones of children as direct result of nuclear bomb tests. In the early 1960s, Linus Pauling challenged President Kennedy to negotiate the Atmospheric Test Ban Treaty with the Soviet Union to put an end to the dangers that Commoner had so ably exposed in the preceding decade. As scientists, both men had grown concerned, not only about the dangers of nuclear weapons testing, but about the militarization of science that had taken hold during the Second World War and intensified during the early 1950s as the exigencies of Cold War geopolitics had become particularly intense.

Barry Commoner's experience with the Navy during the Second World War had opened his eyes to the danger of the misapplication of new technologies. He participated in an experiment with the aerial dispersal of DDT on a stretch of the New Jersey coastline, in order to assess its effectiveness in wiping out mosquitoes and their attendant threat of malaria in the islands where American forces had been engaged in combat with Japan. Commoner witnessed the incredible power of DDT to eliminate mosquitoes overnight, but he also got a firsthand look and the aftereffects that soon followed. At first the beaches subjected to DDT were free of insects, but within days they were covered with many more insects than had been there before. These pests had arrived to feast on the thousands of fish that had also been killed by the DDT dispersal (Egan 2007, pp. 21–22). From experiences such as this, Commoner formed a conclusion about the application of technology that would be central to his long career as a scientist and political activist: the reckless application of new technologies can pose a grave threat to human health and to the environment.

The conclusion which formed the basis of Commoner's now seems obvious enough to be a truism, but he found that it received little attention during the boom decades of economic growth following the Second World War. Man-made dangers to public health were often ignored in the 1950s because many in the public had come to see large-scale research and development projects as more than competent to govern themselves (Egan 2007, p. 148). After all, such research programs, run by both the military and private industry, had been essential to the Allied victory in World War Two, and to the intense period of economic growth that followed. This sense of public complacency about the development and implementation of new technologies for military and commercial purposes was precisely what alarmed scientists such as Pauling and Commoner and motivated them to take action.

After the success of the ban on the atmospheric testing of nuclear weapons in 1963, Barry Commoner set out to analyze the place of science and technology within the broader context of American society. In *Science and Survival*, which Commoner published in 1966, he argued that transparency and the free exchange of information were essential to good science, but that these values had been severely compromised in the twentieth century, and especially since the Second World War. Although government and commercial funding for scientific research had exploded as a result of World War Two, the Cold War, and the phenomenal economic boom of the fifties and sixties, these sources of funding for science had increasingly compromised its core values. The government's most compelling reason for funding scientific research was to enhance the strength of the United States armed forces, thus compelling an atmosphere of strict secrecy concerning scientific research as well as subjecting scientists themselves to unprecedented scrutiny concerning their political beliefs and even their personal lives. Commercial investment in scientific research came with another set of strings attached. In this case, strict secrecy was required to protect intellectual property and to avoid ceding any advantage whatsoever to competing businesses.

As Commoner saw it, the abrogation of transparency in the practice of science was coupled with a fundamental corruption of its most noble goals, including the desire to understand the cosmos and to discover knowledge that could be useful to all of humankind. Instead of serving these goals, science locked into the service of military or corporate interests could deliver results that were more horrifying than inspiring, such as the Hydrogen bombs exploded by the U.S. and Soviet Union in the early 1950s, or the morning sickness drug Thalidomide that was marketed by a West German pharmaceutical firm in the second half of that decade – and which soon proved to cause serious birth defects. In the case of each technology, public awareness of the dangers involved was obstructed by the cultures of governmental or corporate secrecy that Commoner identified as inimical to science itself (Egan 2007, p. 18).

In addition to criticizing the distortion of science by governmental and corporate secrecy, Commoner was also critical of the tendency of Neo-Malthusian thinkers such as Garrett Hardin and Paul Ehrlich to place the values of human rights and environmental sustainability in opposition to each other. In response to Ehrlich's claim that the only effective way to reduce pollution was to stop human population growth, Commoner argued that such a strategy was, "equivalent to attempting to save a leaking ship by lightening the load and forcing passengers overboard. One is constrained to ask if there isn't something radically wrong with the ship" (Egan 2007, p. 126). Like Linus Pauling, Commoner argued that measures to fix the global ship of state should address *both* human rights and environmental sustainability, instead of sacrificing one of these values in the name of advancing the other. Perceiving a symbiotic relationship between human rights and ecology, Commoner struggled to protect both throughout his long career as a scientist and activist.

By the mid-1960s public awareness of environmental problems was on the rise. In some part, this was due to Pauling and Commoner's campaign against the atmospheric testing of nuclear weapons, but for the most part the new sense of environmental awareness emerged in response to the publication of Rachel Carson's *Silent Spring*. An instant best seller, Carson's book was published just before the Cuban Missile Crisis in the anxious autumn of 1962. Although primarily focused on the overuse of insecticides and their devastating effects on the earth's ecosystems, *Silent Spring* also called into questions current assumptions about the practice of science. In the book's last chapter, Carson took aim at the idea that our scientific knowledge could ever endow us with the power to control nature. Carson advocated biological solutions to pest control, such as the introduction of sterilized males to mosquito populations. Although she was an accomplished marine biologist who had worked for years at Woods Hole Oceanographic Institution and written critically acclaimed bestsellers explaining the principles of marine biology to the general public, Carson's arguments in Silent Spring were dismissed by her opponents such as the chemist and American Cyanamid spokesman Robert White-Stevens as being anti-scientific. Where Carson had argued that it was an absurd fantasy to believe that humankind could control nature, White-Stevens argued that our species control of nature was an established fact, and to reject that fact was itself unscientific. Although White-Stevens attempted to brand Carson as a "hysterical" woman possessed by an irrational fear of science and progress, she was, like Commoner and Pauling, steadfastly committed to promoting public policy based on the best science available (Lear 1997, pp. 447–450).

Since the publication of *Silent Spring*, Carson's strategy of informing the public about relevant developments in science and technology and working to maintain standards of transparency and accountability from governments and private corporations has produced a number of important reforms that have helped to protect both the environment and public health. The Atmospheric Test Ban Treaty of 1963, the U.S. ban on the domestic use of DDT, the dramatic curtailment of civilian nuclear power after the Three Mile Island accident in 1979, and stricter government regulation of the tobacco industry are all examples of how greater demands for accountability and transparency have successfully challenged powerful political and economic interests in order to protect public health and the environment. If agitation by scientists and citizens in the twenty-first century leads to a genuine reduction in greenhouse gas emissions across the world, that success would illustrate the symbiotic relationship between science and democracy on a global scale.

In his 1963 Nobel Prize Address, Linus Pauling alluded to the same dream of global democracy articulated by Einstein at the dawn of the nuclear age. When accepting the prestigious award for his work on advancing the Atmospheric Test Ban Treaty, Pauling declared that the only hope of preventing nuclear war and annihilation lay in the gradual establishment of "world law" based on the principles of democracy. Aware that the United States government had actively subverted democratically elected governments as part of its Cold War geopolitical strategy, Pauling argued that such interventions presented a serious obstacle to the development of those global standards necessary to achieve peace and survival in the nuclear age:

> It may take many years to achieve such an addition to the body of world law. In the meantime, much could be done through a change in the policies of the great nations. During recent years, insurrections and civil wars in small countries have been instigated and aggravated by the great powers, which have moreover provided weapons and military advisers, increasing the savagery of the wars and the suffering of the people. In four countries during 1963 and several others during preceding years, democratically elected governments, with policies in the direction of social and economic reform, have been overthrown and replaced by military dictatorship, with the approval, if not at the instigation of one or more of the great powers. These actions of the great powers are associated with policies of militarism and national economic interest that are now antiquated. I hope that the pressure of world opinion will soon cause them to be abandoned and to be replaced by policies that are compatible with the principles of morality, justice, and world brotherhood. (Pauling, Nobel Lecture. December 11, 1963. URL: http://www.nobelprize.org/nobel_prizes/peace/laureates/1962/pauling-lecture.html)

The concept of a broadly democratic framework of "world law" in which "the pressure of world opinion" would check the more atavistic practices of great powers was not unique to Linus Pauling, nor to scientists from the Western side of the Cold War divide. Before the end of the 1960s, the Soviet physicist Andre Sakharov would soon issue his own calls for greater transparency and accountability as essential to human survival (Gorelik and Bouis 2005.pp 275–277).

The challenges that Pauling had faced in advancing the cause of the Atmospheric Test Ban treaty point to a fundamental tension between democracy and diplomacy that will probably always exist, but which has been somewhat ameliorated through the evolution of international forums over the past several generations. Democracy requires a high degree of transparency in order to be democratic. Diplomacy requires a high degree of discretion in order to be diplomatic. Integrating democracy and diplomacy has always been difficult for this reason, but a century of experience tells us that it is not impossible. Woodrow Wilson attempted to minimize secrecy in international diplomacy after the First World War by advocating the creation of the League of Nations as a forum for the creation of "open covenants, openly arrived at." Although the League failed to prevent the Second World War, its existence exposed the causes of that war and brought the regimes that started it to the light of day. International scrutiny of national policies resumed again after the Second World War, thanks not only to the United Nations, but also to the growth of electronic media such as satellite television and the Internet. These trends have exposed the environmental records of industrialized nations to a degree of international scrutiny that they would not otherwise face.

Democratic institutions such as a free press, independent courts, and fair elections are as essential to the health of a society as a functioning nervous system is to the health of an individual. In Love Canal, New York, for example, the citizen activist Lois Gibbs compelled the federal government of the United States in 1978 to take notice of a toxic waste problem that it had previously ignored. The scandal of the toxic waste dump at Love Canal raised concerns across the entire country and led to the passage of the Comprehensive Environmental Response, Compensation, and Liability Act of 1980, which charged polluters and created a Superfund to clean up toxic waste sites throughout the United States. Without a free press, the story of Love Canal would have remained buried, and, without a democratically elected legislature and executive, the Superfund program would never have been created. Lois Gibbs, who began her activism as a concerned mother of two children suffering from the health effects of toxic waste exposure, went on to found the Center for Health, Environment & Justice, an organization dedicated to "empowering people to build healthy communities, and preventing harm to human health caused by exposure to environmental threats" (CHEJ website http://chej.org/about/mission/ Accessed June 23, 2014). Although she had not been trained as a scientist before exposing the problems at Love Canal, Gibbs created an organization that has coordinated its work with environmental scientists, such as Barry Commoner, to address public health problems facing ordinary citizens. If we are to meet the challange of climate change, the democratic institutions that have made this sort of work possible need to be protected, expanded, and implemented on a supranational scale. The prospect of supranational democracy, and the questions that it raises, will be the subject of the next chapter.

## Bibliography

Anderson, Benedict. 1983. *Imagined communities: Reflections on the origin and spread of nationalism*. London: Verso.

Bess, Michael. 2006. *Choices under fire: Moral dimensions of world war two*. New York: Vintage Books.

Carlarne, Cinnamon P. 2010. *Climate change law and policy: EU and US approaches*. Oxford: Oxford University Press.
Center for Health, Environment & Justice website. 2016. Accessed 23 June 2014. http://chej.org/about/mission/
Deese, R.S. 2015. *We are amphibians: Julian and Aldous Huxley on the future of our species*. Oakland: University of California Press.
Egan, Michael. 2007. *Barry commoner and the science of survival: The remaking of American environmentalism*. Cambridge: MIT Press.
Fang, Lizhi. 1989 "*Keeping the faith*" New York Review of Books. December 21st, 1989.
Ferris, T. 2011. *The science of liberty: Democracy, reason, and the laws of nature*. New york: Harper Perennial.
Gorelik, Gennady, and Antonina W. Bouis. 2005. *The world of Andrei Sakharov: A Russian physicist's path to freedom*. Oxford: Oxford University Press.
Hansen, James. 2009. *Storms of my grandchildren: The truth about the coming climate catastrophe and our last chance to save humanity*. New York: Bloomsbury.
Hesiod. 2017. *The poems of Hesiod: Theogony, works and days, and the shield of Herakles*. Trans. by Barry B. Powell. Oakland: University of California Press.
Holden, Barry, ed. 2000. *Global democracy: Key debates*. London: Routledge.
Ibsen, Henrik. 1882. Trans. Nicholas Rudall. (2007) An enemy of the people. Chicago: Ivan R. Dee.
Kant, Immanuel [1795] 1983. Trans. Ted Humphrey. *Perpetual peace and other essays*. Indianapolis: Hackett Publishing.
Kasparov, Garry. 2015. *Winter is coming: Why Vladimir Putin and the enemies of the free world must be stopped*. New York: Public Affairs Books.
———. 2017. "Donald's Pravda: Trump and his apologists spookily echo Vladimir Putin" *New York Daily News*. Sunday, July 16, 2017.
Koehn, P.H. 2015. *China confronts climate change: A bottom-up perspective*. New York: Routledge.
Krasilovsky, Alexis. 2017. *Great adaptations: screenwriting and global storytelling*. New York: Routledge.
Lear, Linda. 1997. *Rachel carson: Witness for nature*. New York: Henry Holt & Company.
MacAskill, Ewen. 2008. "Hurricane Gustav: Republican Convention Thrown into Chaos" *The Guardian*, 31 August, 2008.
Mirsky, Steven. 2017. "Dirty doctors finished what an Assassin's bullet started: Disregarding new scientific information can be deadly" *Scientific American*. February 1st, 2017.
Monnet, Jean., Richard Mayne, trans. 1978. Memoirs. New York: Doubleday & Company.
Open Secrets. 2018. Campaign contribution profile for Senator James Inhofe. https://www.opensecrets.org/members-of-congress/summary?cid=N00005582&cycle=CAREER (Accessed 14 Jan 2018).
Oreskes, Naomi, and Erik M. Conway. 2011. *Merchants of doubt: How a handful of scientists obscured the truth on issues from tobacco smoke to global warming*. New York: Bloomsbury.
Pauling, Linus. December 1963."*Science and peace*." Nobel lecture. http://www.nobelprize.org/nobel_prizes/peace/laureates/1962/pauling-lecture.html
Purdy, Jedediah. 2015. *"Environmentalism's Racist History" The New Yorker.* August 13th, 2015.
Ricks, Thomas. 2017. *Churchill and Orwell: The fight for freedom*. New York: Penguin Press.
Riesbeck, David J. 2016. *Aristotle on political community*. Cambridge: Cambridge University Press.
Runciman, David. 2013. *The confidence trap: A history of democracy in crisis from world war I to the present*. Princeton: Princeton University Press.
Shapiro, Judith. 2001. *Mao's war against nature: Politics and the environment in revolutionary China*. Cambridge: Cambridge University Press.
Standage, Tom. 1998. *The victorian internet: The remarkable story of the telegraph and the nineteenth century's online pioneers*. New York: Bloomsbury.

Tal, Alon, ed. 2006. *Speaking of earth: Environmental Speeches that moved the world*. New Jersey: Rutgers University Press.

Weart, Spencer. 1998. *Never at war: Why democracies will not fight one another*. New Haven and London: Yale University Press.

World Health Organization. 2005. *Chernobyl: The true scale of the accident*. Geneva: Joint News Release WHO/IAEA/UNDP. September 5th, 2005. http://www.who.int/mediacentre/news/releases/2005/pr38/en/.

# Chapter 7
# Supranational Democracy

*"If you think in terms of people divided up into countries, you won't follow me. The idea of countries is going by the boards. Young people are getting wonderfully uprooted and they're too strong to get sucked into this 'country' crap."*

Buckminster Fuller

**Abstract** Advocates of expanding democracy beyond the current Westphalian system of sovereign nation states can find some valuable precedents in the history of cosmopolitan political thought from ancient times to the present. Within this long history, the proposals for constructing a democratic federation on a world scale that emerged in response to the advent of nuclear weapons in the mid-twentieth century deserve special examination. In light of recent advances in communication technology, some of these proposals have become more plausible, and may offer new tools for addressing the problem of climate change in the twenty-first century.

**Keywords** Mozi · Cicero · de Vitoria · Grotius · Kant · Cosmopolitan democracy

In light of the growing need for democratic accountability to protect the global commons, the case for creating democratic institutions on a global scale is stronger than it has ever been before. If we want to extend the environmental benefits of democracy to global governance, we need to create a cosmopolitan democracy. It is necessary to acknowledge, however, that "cosmopolitan" is a word that carries a certain amount of baggage. If we don't think of the magazine or the cocktail when we hear the word cosmopolitan, we are likely to think of an ideological worldview that took a very serious beating from both the right and the left in the twentieth century. For those on the right, a "rootless cosmopolitan" was one who lacked loyalty to the nation and its values; in the hands of Fascists, the term became a slur for those, such as European Jews or the Romani people, who would soon be the targets of systematic persecution and genocide. On the left, however, cosmopolitanism also came under attack in the twentieth century. Following Lenin's 1916 declaration that



R. S. Deese, *Climate Change and the Future of Democracy*, Environmental Challenges and Solutions 5, https://doi.org/10.1007/978-3-319-98307-3_7

imperialism was the highest stage of capitalism, it became the policy of Marxist-Leninist movements to cultivate nationalist uprisings against colonial empires throughout the world. When Lenin seized power from the provisional government of the liberal democrat Alexander Karensky in November of 1917 he would soon be able to back his idea with the resources of the Russian state. While some leading leftists, such as Rosa Luxemburg, condemned Lenin for establishing an authoritarian regime that would "inevitably cause a brutalization of public life," other leftists saw Leninism, especially after the Bolsheviks consolidated their national power in 1922, as the wave of the future (Kaplan 2008, p. 72). Subsequently, many on the left embraced the Marxist-Leninist revolutionary movements that emerged across the world in the twentieth century. Often arising in response to the brutality of colonial rule in Asia, Africa, and Latin America, the Marxist-Leninist revolutions that proliferated after 1945 embraced an ethos of intense nationalism, and often saw cosmopolitanism as a stalking horse for Western imperialism. Even among intellectuals in Western nations, cosmopolitan ideals lost much of their luster in the twentieth century. With the rise of postmodern critiques of the Enlightenment in the closing decades of the twentieth century, the cosmopolitan vision of modernity fell into further disrepute in academia. In the somewhat more conservative field of International Relations, cosmopolitanism also fell out of fashion after World War Two, as leading intellectuals such as George F. Kennan and Reinhold Niebuhr dismissed efforts to establish democratic global governance as both naïve and dangerous.

Though Kennan had never been drawn to world federalism, Niebuhr had for a short time been involved with the Committee to Draft a World Constitution at the University of Chicago, and so based his critique of the movement to some extent on his personal experience. The constitution that was drafted in Chicago envisioned a world parliament composed of units based on both geography and culture. For example, it left open the question of whether the United Kingdom would be part of the European regional or whether it would join Canada and the United States as part of a transatlantic bloc. The question of electoral representation presented an even greater conundrum to the committee. Niebuhr recounted how the proposal that western nations receive greater representation than the more populous nations of Asia was supported by some world federalists, such as the physicist Harold Urey, but rejected as patently "immoral" by others, such as Colonel M. Thomas Tchou from the Republic of China. Recounting this tense exchange, Niebuhr reflected wryly that "real problems have an inconvenient habit of peeking through, even at a dinner of a World Republic convention" (1949, p. 386). Incidents such as this shed light on the crux of Niebuhr's argument against world federalism. He reasoned that a democratic government required a reasonably cohesive community for its foundation, and that the vastly different nations of the world in the middle of the twentieth century were nowhere close to comprising such a community.

There is little doubt that revolutions in transport, commerce, and communication have made the world a much smaller place than it was in 1949, when Niebuhr penned his critique of the world federalist movement. Nonetheless, the sense that regional and cultural blocs have radically different goals and values remains a formidable obstacle to any attempt create a democratic government on a global scale.

In the aftermath of the Cold War, when liberal democracy seemed to be ascendant across the world, Samuel Huntington's influential book *The Clash of Civilizations and the Remaking of World Order* argued that dangerous fault lines still divided the cultural regions of the world, and conflicts between the Judeo-Christian nations of the West and the Islamic societies that spanned vast swaths of Asia and Africa were likely to persist for a long time to come (Schweickart 2011, p. 213). First published in 1996, Huntington's bestseller makes the case that the world is divided into distinct civilizations defined primarily by religion, and that these civilizations a locked into a continuing dynamic of conflict and competition. Not surprisingly, the events of September 11th, 2001 were seen by many as a confirmation of Huntington's worldview, and cast further doubt on the idea that the world could ever be united by the universal acceptance of liberal democracy and free market capitalism.

While the religious divisions that Huntington described have had a real impact on world events, the is considerable evidence that his schema overstates the influence of religion to the exclusion of other factors. Most glaringly, such a paradigm ignores the considerable division and conflict *within* the alleged "civilizations" of the world. In the case of the Islamic world, for example, the divisions between Sunni and Shia Muslims are often a greater source of violence than the divisions between Muslims and the confessional communities of other major religions. Furthermore, the divisions between those who take a fundamentalist approach to religion and those who do not has proven to be a powerful source of cultural conflict within many of the world's religious communities, including Islam, Christianity, Judaism, and Hinduism. In his book *Identity and Violence*, the economist and philosopher Amartya Sen effectively dismantled Huntington's notion of "civilization" by illuminating the myriad factors, including "nationalities, locations, classes, occupations, social status, language, [and] politics. .." that create "other systems of partitioning" within the religious communities of the world (2006, p. 10). Sen argues that whether one is warning about a "clash of civilizations" or calling for a "dialogue among civilizations," there is a fundamental flaw in any paradigm that ignores the irreducible complexity of human individuals by lumping them into such an archaic and inflexible category. While Sen concedes the considerable influence of both religion and traditional culture in world affairs, he cautions that the conception of the world as a group of civilizations "speedily reduces many-sided human beings into one dimension" and ignores "the variety of involvements that have provided rich and diverse grounds for cross-border interactions over many centuries" such as "arts, literature, science, mathematics, games, trade, politics, and other arenas of shared human interest" (2006, p. 12).

Among those "other arenas of shared human interest" the desire to protect future generations from the dangers posed by catastrophic climate change and to ensure their survival and wellbeing must certainly hold a privileged place. As James R. Huntley argues in his conclusion to *Pax Democratica*, the goal of protecting free elections and universal human rights represents a cause capable of inspiring hard work and sacrifice from all corners of the world: "Such world embracing principles—and not some spurious new super-nationalism or devotion to a 'civilization'—represent a good enough creed on which to found" an intercontinental union of the world's democracies in the twenty-first century (Huntley, 2001. p. 189). Nonetheless, to imagine that the creation of a global polity is possible is not that same as to accept that it is desirable. The fear remains that a government of such scale, even if it were thoroughly democratic to begin with, could devolve into something inimical to human freedom. To be worried about the overreaching power of a single unified state is to be in very good company. Erasmus worried that the unification of Europe could lead to tyranny, and Jefferson voiced a similar concern about the rise of a stronger federal government in the United States following the constitutional convention of 1787. The dream of a technocratic world government promoted by H.G. Wells was highly disturbing to Aldous Huxley and the dystopian depiction of the World State that he sketched in *Brave New World* was aimed directly at the ideas that Wells had been promoting for the first three decades of the twentieth century. The dystopian genre that Aldous Huxley helped to launch has solidified popular fears of world government as something that would most likely destroy the soul of the human race. Because a global democracy could theoretically devolve into a global tyranny, there seems to be a strong liberal case for caution about the concept of supranational governance, even if its foundations are democratic. As the philosopher Kwame Anthony Appiah put it, "A global state . . . could easily accumulate uncontrollable power, which it might use to do great harm" (Appiah 2006, p. 163).

Whether we say it out loud or not, a compelling syllogism seems to haunt our thoughts and to abort most conversations about the idea of global democracy before they can even begin. With minor variations, it goes as follows: *Global democracy is a form of global government. Any form of global government is bound to become a soul-crushing dystopia. Ergo, advocates of global democracy are really (whether they know it or not) advocates of a soul-crushing dystopia.* However, liberals who fear creeping tyranny must acknowledge that the greatest threats to individual liberty have not come from supranational organizations such as the EU or UN, but from the prerogatives that sovereign nation states routinely exercise in the name of national security. While supranational institutions have acquired fairly limited powers since 1945, the extra-constitutional powers acquired by agencies tasked with protecting *national* security have multiplied with every crisis, from World War Two to the Cold War to the War on Terror.

In the late 1930s, Clarence Streit made the argument that the sovereign nation-state was a greater threat to individual liberty than his proposal for a global union of democracies. Surveying contemporary reactions to his proposal, Streit observed that many advocates of democracy feared that a union of democracies would constitute a new kind of "super-state" that would threaten liberty. For Streit, such warnings against integrating the world's democracies were predicated on the false assumption that "territory meant tyranny." Countering this assumption, Streit pointed out that, historically, "the states that gave us the word 'tyrant' were among the smallest, not the biggest" and that democracy became stronger in Switzerland when the various cantons combined into an Alpine "super-state" to protect their common freedom. Streit concluded that, "it is not size which the individual really fears in the state, but power over himself, interference in his liberties, meddling in his life." In all of these areas, the sovereign nation state, in jealous economic and strategic competition with its rivals, presented a greater danger than a union of democracies would (Streit 1939, pp. 187–188). Nonetheless, the concept of anything like a world state had taken on sinister connotations in the popular imagination by the time Streit published *Union Now*, as evidenced by the success of Huxley's *Brave New World* after its publication in 1932.

In light of Aldous Huxley's influence on cultural perceptions of global government, it is worth remembering that his vision of a repressive World State is not the only blueprint for dystopia nor even the likeliest. Huxley's student at Eton, Eric Blair, would come to share his teacher's disdain for H.G. Wells and his vision of a technocratic state. Taking the pen name George Orwell, he would express his admiration for Huxley's dystopia and, immediately following World War Two, set out to construct his own. However, the dystopian hell depicted in *Nineteen Eighty-four* is not the creation of a single world state, but rather the product of several competing empires (Oceana, Eurasia, and Eastasia) in a state of perpetual war. As a seasoned war correspondent and acutely perceptive political thinker, George Orwell understood that war was the ideal vehicle for enhancing state power, and would likely remain so for a very long time to come.

If we examine where totalitarianism flourished in the twentieth century, Orwell's instinct about the natural affinity between perpetual war and perpetual tyranny is well confirmed. From Lenin's repressive "War Communism" of the early 1920s to Stalin's man-made famines and political purges the following decade, the growth of state power in the Soviet Union was justified by a steady stream of military confrontations with foreign and domestic enemies. As Mussolini and Hitler built their respective totalitarian states in Italy and Germany, they borrowed some tactics from the Soviet system, but made a relentless war against international communism their primary justification for expanding state power and pulverizing individual rights. In the United States, the exigencies of war, both hot and cold, would lead to the worst violations of constitutional rights in the twentieth century, from the persecution of political dissidents during the First World War, to the internment of Americans of Japanese ancestry in the Second World War, to the anticommunist witch hunts of the Cold War era. In the twenty-first century, the events following September 11, 2001 have only added further example of the synergetic relationship between perpetual war and the expansion of state power. In 1917, the leftist intellectual Randolph Bourne could not have known how thoroughly the century ahead would vindicate his observation that "War is the health of the state" (Clayton, 1998, p. 252.)

In light of the affinity between war and the metastasis of state power, it seems more likely than not that any system in which sovereign nation states reserve the unconditional right to wage war on their own terms would increase the probability of tyranny *within* nation states. The answer to this problem, as Immanuel Kant saw it, was not the elimination of nation states themselves, but a federation of free states (which Kant called "republics") that would minimize the possibility of interstate conflict. In his essay *On Perpetual Peace*, he imagined a European system that he hoped would also ensure a republican form of government among its sovereign states. Because the citizens of any state pay the price in blood and treasure for war,

Kant reasoned that states that remained accountable to their citizens would be more likely to avoid war. Kant recognized that the relationship between autocracy and warfare can easily devolve into a vicious cycle: the exigencies of war makes the state less accountable to its citizenry, and this loss of accountability in turn makes the state more warlike. Although he was writing in the last decade of the eighteenth century, Kant also seemed to anticipate the horrific scale of warfare that would emerge in the twentieth century. Condemning the extermination of the enemy as a goal in warfare, Kant seemed to anticipate the concept of Mutually Assured Destruction that would emerge as a consequence of the nuclear arms race. Kant reasoned that, "a war of extermination, in which both parties. . . can be eradicated simultaneously could bring about perpetual peace only over the great graveyard of humanity. Such a war, therefore, and hence the use of the means that would lead to it, must be utterly forbidden" (Kant [1795], 2006. p. 71). Kant's words anticipate those of President Dwight Eisenhower, when he contemplated the economic and social consequences of the nuclear arms race in the early 1950s:

> This world in arms is not spending money alone. It is spending the sweat of its laborers, the genius of its scientists, the hopes of its children. The cost of one modern heavy bomber is this: a modern brick school in more than 30 cities. It is two electric power plants, each serving a town of 60,000 population. It is two fine, fully equipped hospitals. It is some fifty miles of concrete pavement. We pay for a single fighter plane with a half million bushels of wheat. We pay for a single destroyer with new homes that could have housed more than 8,000 people . . . This is not a way of life at all, in any true sense. Under the cloud of threatening war, it is humanity hanging from a cross of iron. (Eisenhower 1953)

To contemplate Kant's musings from 1795 about the "graveyard of the human race" alongside Eisenhower's lamentations in the spring of 1953 about "humanity hanging from a cross of iron" inspires a sense of awe about the ability of the human race to make enormous progress in some fields of endeavor, while remaining seemingly paralyzed in others. On the face of it, it would seem that the problem is one of technical progress accompanied by moral paralysis, as illustrated by the fact that humanity created new weapons with the power to exterminate millions of souls in the 20th century, without creating institutions powerful enough to control those weapons. It has been more than two centuries after Kant declared that, "a war of extermination" and "the means that would lead to it must be utterly forbidden." Nonetheless, there are still no supranational institutions powerful enough to effectively prevent, with the threat of certain punishment, any state that pursues such a war. While the United Nations was founded on the pretense that it could be that sort of institution, every single permanent member of its Security Council has sought to protect its own security by producing and maintaining an arsenal of nuclear and thermonuclear warheads; i.e. weapons that could only be used for the very type of genocidal warfare that the United Nations was ostensibly founded to ban forever. In spite of the legitimate reservations that many have expressed about any form of cosmopolitan government, we have to weigh those reservations alongside the dangers posed by a world of competing sovereign nation states.

Although Kant was perhaps its most famous exponent among Enlightenment philosophers, the idea of a cosmopolitan government for the benefit of all people has a long and illustrious history. In the warring states period, the Chinese philosopher Mozi advocated pacifism and universal love as the only sound basis for government. Mozi decried the destructiveness of war, and pointed to the unifying concept of heaven:

> But how do we know that Heaven loves all the people in the world? Because it enlightens them all. How do we know that it enlightens them all? Because it possesses them all. How do we know that it possesses them all? Because it feeds them all. I say: within the four seas (the world) all grain-eating (civilized) people feed oxen and sheep with grass and dogs and pigs with grain, and cleanly prepare pastry and wine to sacrifice to the Lord on High and spiritual beings. Possessing all people, how could Heaven not love them? (Chan 1963, p. 227).

In Chinese culture, the emperor was conceived of as the "son of heaven" and many local princes dreamed of conquering their neighbors and establishing themselves as universal rulers legitimated by the "mandate of heaven." However, Mozi argued that the military path to power could not be reconciled with the mandate of heaven. He reasoned that, "to recruit the people of Heaven to attack the cities of Heaven is to murder the people of Heaven, smash altars, demolish shrines, and kill sacrificial animals. In this way, on the higher level no benefit to Heaven can be attained" (Chan 1963, p. 227). Mozi is commemorated as a significant figure in the history of Chinese philosophy, but it may not be surprising that his thinking was never embraced officially. The Qin, who unified China, embraced the brutally efficient philosophy of Legalism, which emphasized the maintenance of state power through fear. The Han, who established the first examinations that would characterize imperial rule in China until the early twentieth century, rejected both Legalism and the philosophy of Mozi in favor of Confucianism.

In ancient Greece, Socrates may have been the first philosopher to describe himself as a "citizen of the world" though in the end he chose death over leaving his beloved city state of Athens. Diogenes of Sinope, on the other hand, not only proclaimed his lack of allegiance to any city state but demonstrated it relentlessly by destroying money, defying every norm of custom, and showing no deference at all to wealth and power. Plutarch described an encounter between Diogenes and the emperor Alexander the Great, that would be celebrated in art and literature for centuries to come:

> Alexander went in person to see [Diogenes]; and he found him lying in the sun. Diogenes raised himself up a little when he saw so many people coming towards him, and fixed his eyes upon Alexander. And when that monarch addressed him with greetings, and asked if he wanted anything, "Yes," said Diogenes, "stand a little out of my sun." It is said that Alexander was so struck by this, and admired so much the haughtiness and grandeur of the man who had nothing but scorn for him, that he said to his followers, who were laughing and jesting about the philosopher as they went away, "But truly, if I were not Alexander, I would be Diogenes." (Hamilton 1999, p. 14).

The school of ancient philosophy that did the most to advance a disciplined and exemplary cosmopolitan view of the world was Stoicism, founded by Zeno of Citium. Although Stoicism began in Greece, it was ultimately promoted by prominent Romans such as the great Senator and statesman Cicero. Convinced that republican government was the highest form of civilization, Cicero believed that justice and law transcended geography and nationality:

> There is a true law, a right reason, conformable to nature, universal, unchangeable, eternal, whose commands urge us to duty, and whose prohibitions restrain us from evil. . . . It is not one thing at Rome and another at Athens; one thing today and another tomorrow; but in all times and nations this universal law must forever reign, eternal and imperishable. (Owen and Owen 2010, p. 223)

Before the Roman republic devolved into an empire, Cicero expressed his conception of a *res publica totius orbis*, or "the republic of the whole world" (Talbott 2009, p. 47). Cicero's cosmopolitan vision of the Roman republic demanded a sense of universal law, a strong ethos of personal virtue, and a division of powers among consuls to check individual ambition and preserve the rule of law. While these values were occluded by Rome's shift to empire, and its ultimate decline, they left a legacy that would later prove important to European political thought.

When Spain and Portugal emerged as global empires in the sixteenth century, the cleric and scholar Francisco de Vitoria revived Cicero's concept of *res publica totius orbis* to argue that the native peoples of the newly discovered western hemisphere had some legal standing that could not be dismissed in the name of conquest. Echoing the expansive vision of the great Roman senator, he argued:

> The whole world, which is in a sense a commonwealth, has the power to enact laws which are just and convenient to all men; and these make up the law of nations. From this it follows that those who break the law of nations, whether in peace or in war, are committing mortal crimes, at any rate in the case of the graver transgressions such as violating the immunity of ambassadors. No kingdom may choose to ignore this law of nations because it has the sanction of the whole world. (de Vitoria 1991, p. 40)

Francisco de Vitoria's lectures at the University of Salamanca were widely discussed among educated people in Spain, and they came to have a profound influence on Bartolomé de las Casas, who struggled to end the worst abuses against Native Americans at the hands of the Spanish conquistadores (Scott 2000, p. 77).

The Salamanca school of thought advanced by Francisco de Vitoria would soon influence the thinking of the Dutch jurist Hugo Grotius. Countering claims made by Portugal that it alone possessed the right of free navigation in the Indian Ocean, Grotius hoped to promote Dutch interests in the East Indies by advancing the idea that all nations had an equal right to navigate the open seas for the purpose of trade. His treatise on this subject, *Mare Liberum*, would become one of the foundational documents in the field of international law. Grotius made his argument in favor of free trade, but it had implications beyond trade: "The following most specific and unimpeachable axiom of the Law of Nations, called a primary rule or first principle, the spirit of which is self-evident and immutable, to wit: Every nation is free to travel to every other nation, and to trade with it" (Grotius, 1609). Although he was writing two centuries before the utilitarian philosopher Jeremy Bentham would coin the term "international," the concept of the "law of nations" that Grotius used here implies that there must be a governing set of norms to make trade among nations both possible and profitable.

In the eighteenth century, the concept of federalism would become an important element in the political discourse of the Enlightenment. The root word of federal is the Latin verb *fidere*, meaning to trust. Strobe Talbott has observed that "the word *federal* had previously been used most commonly as a theological term that deferred to the divine devolution of responsibility to human beings." In David Hume's lifetime, the term began to acquire a more secular meaning, as it described any political system "in which the highest authority devolved power to subsidiary ones." Hume saw the potential value of political federalism if it could be refined to serve the common good while leaving local communities with a high degree of autonomy over their own affairs. Talbott describes Hume's vision of federalism as one that would "combine the advantages of local accountability with those of a large state" by leaving "as much power as possible at the local lower levels while investing as much as necessary in the higher ones" (Talbott 2009, p. 92).

Benjamin Franklin would be one of the first to introduce the concept of federalism into the political discourse of the Anglo-American colonies. In response to the crisis of the French and Indian Wars, Franklin proposed one of the most ambitious schemes for political federation of his time when he put forward the Albany Plan of 1754, which would have entailed a formal political union of the American colonies and the creation of a council appointed by elected representatives from each of the colonies. Although conceived with a deference to monarchy that would have left the lion's share of political authority with the British crown, Franklin's Albany Plan proposed the creation of a federal system of elected representatives that was unprecedented in its geographical scale. The boldness of Franklin's vision was generously confirmed by the vehemence with which both the monarchy and his fellow American colonists rejected it. To the crown, the creation of a unified government for the colonies ran counter to the ancient practice of empire, divide and rule. For the colonial governors up and down the eastern seaboard, the idea of political unification with other colonies was a threat to their local power. As in so many other respects, Franklin saw farther than his contemporaries. Two decades after the Albany Plan was universally rejected, its core idea of an elected council of the colonies was embraced with the creation of the Continental Congress in the autumn of 1774. Less than two years later, that legislative body would initiate the largest experiment in democratic federalism to date with the creation of the United States.

When the United States declared its independence in 1776, Jefferson predicated its legitimacy on the universal rights of "life, liberty, and the pursuit of happiness," and when the French Revolution began in the summer of 1789, the new regime swiftly promulgated its Declaration of the Rights of Man and the Citizen. Both the United States and France would become more nationalistic as their new governments sought to preserve and consolidate their hold on power, but their founding declarations were remarkably universal. As noted earlier, one of the greatest exponents of constitutional democracy and universal human rights in France was the Marquis de Condorcet. When the French Revolution devolved into the Reign of Terror, Condorcet would be condemned to death for his opposition to the death penalty in general and to the execution of Louis XVI in particular. As he awaited death, Condorcet consoled himself by writing a treatise on what he saw as the inevitable progress of the human race toward a new and universal order of liberty and enlightenment, not only for France and America, but for the whole world as well:

> How consoling for the philosopher who laments the errors, the crimes, the injustices which still pollute the earth and of which he is often the victim is this view of the human race, emancipated from its shackles, released from the empire of fate and from the enemies of its progress, advancing with a firm and sure step along the path of truth, virtue, and happiness! It is the contemplation of this project that rewards him for his efforts to assist the progress of reason and the defense of liberty. (Condorcet 1793, p. 201).

The sheer optimism of Condorcet's vision would go on to have a great influence on the utopian socialist thinking of William Godwin in Britain, and Saint Simone and August Comte in France. Regarding the establishment of a federal system of government, however, Condorcet was less influential.

As noted earlier, the most prominent Enlightenment philosopher to systemically link a broad cosmopolitan vision with a commitment to democratic federalism was Immanuel Kant. He did this in two texts written in the late eighteenth century: a short essay called "The Idea for a Universal History with a Cosmopolitan Purpose" published in 1784 and "Toward Perpetual Peace" published in 1795. In "Toward Perpetual Peace," Kant contemplated the possibility that human beings could develop a cosmopolitan ethos, fostered by economic and social interdependence, that would ultimately eliminate war: "The growing prevalence of a (narrower or wider) community among the peoples of the earth has now reached a point at which the violation of right at any *one* place earth is felt in *all* places." Kant therefore argued that "the idea of cosmopolitan right is no fantastic or exaggerated conception of right." (Kant [1795] 2006. p. 84, italics in original).

In contrast to Condorcet, Kant cautioned against believing in the complete perfectibility of the human race, and this essay contains the statement that the twentieth century Isaiah Berlin later celebrated for its mordant accuracy: "Out of the crooked timber of humanity, no straight thing was ever made" (Berlin 2013, p. 50). In his longer treatise "Toward Perpetual Peace" Kant sketched his cosmopolitan vision in greater detail. Arguing that people would most likely retain their national and local loyalties, Kant dismissed the idea of a single world state. What he did find plausible was a federation of republics, governed democratically and linked together in a system of collective security. Kant reasoned that since it is the people who bear the terrible cost of war, a society in which the people were sovereign would avoid war.

In the early 1840s, the seminal Victorian poet Alfred Tennyson advanced a poetic vision of cosmopolitan democracy that was in some ways similar to Kant's. In several stanzas from the poem "Locksley Hall" the poet lets his imagination run towards the distant future. Inspired by recent advances in manned balloon flights, Tennyson imagined a future in which manned flight would first advance trade, and then unleash terrible warfare from the sky. The end result of this he surmised, would be the creation of a global parliament:

> *Till the war-drum throbb'd no longer, and the battle-flags were furl'd*
> *In the Parliament of man, the Federation of the world.*
> *There the common sense of most shall hold a fretful realm in awe,*
> *And the kindly earth shall slumber, lapt in universal law.*
> (Tennyson 1842)

This bold vision appealed to many who were born in the nineteenth century and lived to see the terrible wars of the twentieth century. One of those individuals was Harry Truman, who kept a hand-copied manuscript of "Locksley Hall" in his wallet when he served in France during World War One.

In the early 1940s, as the United States was drawn into the vortex of another world war, the idea of a new international order based on democracy found a growing number of adherents among both Democrats and Republicans. Wendell Willkie, who had been the Republican nominee for President in the 1940 election, published *One World* in 1942, urging the American people to accept membership in a new global organization of states to supplant the failed League of Nations. *One World* was an instant success, which according to the records of its publisher, was the bestselling pamphlet in American history since Tom Paine had published *Common Sense* (Holm 2017, p. 18). When Truman became President in the last months of World War Two, his devotion to Tennyson's vision fueled his commitment to making sure that the UN would succeed where that earlier "Parliament of man," the League of Nations, had failed. In October of 1945, Truman declared:

> It will be just as easy for the nations to get along in a republic of the world as it is for you to get along in the republic of the United States. Now when Kansas and Colorado have a quarrel over the water in the Arkansas River they don't call out the National Guard in each state and go to war over it. They bring a suit in the Supreme Court of the United States and abide by the decision. There isn't a reason in the world why we cannot do that internationally. (*New York Times*. October 10, 1945. p. 20)

In the first months of the postwar era, before the Cold War had come to define U.S. foreign policy, Truman held out hope, at least to the press, that the United Nations could succeed where the League of Nations had failed, and become the foundation for the establishment of "a republic of the world." In practical terms, the three most important international initiatives that Truman and his European allies pursued in the late 1940s were: first, the establishment of the United Nations as a permanent organization to keep peace in the postwar world; second, the prosecution of war crimes at Nuremberg and in Tokyo; and, third, the promulgation of several global declarations and treaties that focused on human rights, such as the Universal Declaration of Human Rights in 1948, the Convention Against Genocide that same year, and the Fourth Geneva Convention in 1949. As the historian Michael Bess has observed, these three initiatives emerged from consensus that rejected the nationalist ethos that had led to World War Two: "Now at war's end, with the imagery of Auschwitz and Hiroshima still fresh in their minds, the world's leaders launched a three-pronged attack on the principle of national sovereignty" (2006, p. 282).

Although many features of the United Nations, such as the veto for permanent members of the Security Council, still protected national sovereignty in matters of war and peace, the need to create new modes of government beyond the boundaries of the nation-state seemed especially urgent after the failure of the League of Nations to prevent World War Two. The primary reason that the League of Nations failed was the absence of the United States, after the U.S. Senate had rejected the

League as a threat to American sovereignty, in spite of the fact that U.S. President Woodrow Wilson, who had cherished the idea of a "parliament of man" since his youth, was one of the chief architects (Kissinger 1994, p. 223). Another major figure behind the idea of a League of Nations was the white South African military leader and politician Jan Smuts (Mazower 2009, p. 192). Although he had fought against the British in the Boer War, he came to find common cause between white South Africans, the British Commonwealth, and the United States as advancing democracy and law across the world. Like Wilson, Smuts was an avowed white supremacist. The white supremacy of leaders such as Wilson and Smuts, in alliance with the Australian delegation, helped to quash a proposal, advanced by Japan at the Versailles conference, that the League Charter include an amendment affirming racial equality (Atkinson 2016, p. 169). Given that anti-colonial struggles were rapidly proliferating in the wake of World War One, the white supremacist politics of the League's founders limited its relevance beyond Europe. As the predatory militarism of autocratic governments in Japan, Italy, and Germany shattered treaties and redrew borders in the 1930s, the League proved to be completely ineffective as system of collective security.

One proposed alternative to the faltering League of Nations was the concept of a full political, economic, and military union of all the democratic countries in the world. The American journalist Clarence Streit proposed this idea on the eve of World War Two, arguing that the institution of the sovereign nation state was becoming inimical to the principles and values of democracy in the modern world. In an especially provocative passage, Streit declared:

> The dictators are right when they blame the democracies for the world's condition, but they are wrong when they blame it on democracy. The anarchy comes from the refusal of the democracies to renounce enough of their national sovereignty to let effective world law and order be set up. But their refusal to do this, their maintenance of the state for its own sake, their readiness to sacrifice the lives and liberties of the citizens rather than the independence of the state—this we know is not democracy. It is the core of absolutism. Democracy has been waning and autocracy waxing, the rights of men lessening and the rights of the state growing everywhere because the leading democracies have themselves led in practicing *beyond* their frontiers autocracy instead of democracy. (Streit 1939, 27. Italics in original).

Just as Tom Paine had attacked the dogma of hereditary monarchy in the late eighteenth century, Streit was now identifying the dogma of national sovereignty as another insidious form of "absolutism" that threatened to thwart the progress of democracy in the twentieth century. For the better part of his long career, Clarence Streit argued for a political, economic, and military union of the world's democracies to counter the threat of fascism. Born in the small town of California, Missouri in 1896, Streit moved to Missoula, Montana as a teenager, and attended the University of Montana until volunteering to serve in the First World War. In the 1920s, Streit became a foreign correspondent for the *New York Times* and covered the early struggles of the League of Nations. By the late 1930s, he became convinced the democratic nations of the world, principally the United States, Britain, and France, needed to form a federal union in order to confront the rising challenges of totalitarianism and militarism. He outlined this vision in his 1939 book *Union*

*Now* and soon founded an organization called the Association to Unite the Democracies to promote greater cooperation, and ultimately a political, economic, and military union, among the world's major democracies.

Clarence Streit's proposal for a "Grand Republic" of the world's democracies was frankly modeled on the success of democratic federalism within his own country (Streit 1939, p. 29). Inspired by the shrewd political genius of Alexander Hamilton and the expansive poetry of Walt Whitman, Streit's vision, though far from parochial, was clearly American in its origins (Stella 2017). This might suggest that his ideas would be a poor fit for the political and economic circumstances of other continents, especially Europe, but this has not been the case. In the late spring of 1940, Winston Churchill stole a page from Streit's manifesto, as he struggled to hold Britain's alliance with France together in the face of the advancing German war machine. On June 16th, Churchill had the British Ambassador to France deliver a proposal for a new political federation in which "France and Great Britain will no longer be two nations but one Franco-British Union." If this proposal were adopted, Churchill promised, "Every citizen of France will enjoy immediate citizenship of Great Britain, every British subject will become a citizen of France" (Reves 1945, p. 249). Though born of Churchill's desperate attempt to keep the French government from capitulating to the Nazis, the idea that Churchill proposed here anticipated some key elements of the process of European integration that would culminate in the European Union before the end of the twentieth century.

During the 1940s, Streit's cause gathered support in the United States from such figures as the Supreme Court Justice Owen J. Roberts, the lawyer and diplomat John Foster Dulles, and the physicist Harold Urey. By the close of that decade, the creation of the North Atlantic Treaty Organization echoed the military component of Streit's vision. As an evangelist for democratic federalism on a global scale, Clarence Streit struck many critics as eccentric in his aspirations and oblivious to the realities of international politics in the twentieth century. In his review of *Union Now*, George Orwell savaged Streit for ignoring the millions of people who were disenfranchised for their skin color when he counted the overseas colonies of Britain, France, and the United States as part of his global union of democracies. George Orwell descried this problem with his characteristic bluntness in his offensively titled essay ("Not Counting Niggers") for the literary journal *The Adelphi* in the summer of 1939. Taking Streit to task for the implicit racism of his "Union of Democracies" Orwell observed that the governments of Britain, France, and the United States ruled over millions of people of color in their African and Asian colonies without their electoral consent. As he saw it, Streit's scheme would do nothing to promote democracy within "the British and French empires, with their six hundred million disenfranchised human beings" (Orwell [1939] 1968. p. 397). Orwell was perhaps unnecessarily offensive in his attack on Streit's myopia in this regard, but there is some justice in his argument that the problems of both colonialism and racism among the Atlantic democracies had not been adequately addressed in Streit's manifesto. Noting Orwell's criticism, the historian Or Rosenboim argues that although "Streit was not a simple apologist for imperialist expansion, he failed to transcend" the broad conceptual categories of "inclusion and exclusion that emerged from imperial history" (Rosenboim 2017. p. 120).

Nonetheless, Streit articulated one of the most fundamental issues at stake in the Second World War when he championed the potential of democratic federalism. Beyond settling the strategic question of which nations would hold sway in Europe and the Pacific during the twentieth century, the outcome of the Second World War also advanced a loose form of federalism as the most widely accepted paradigm in international relations. In some ways, the behavior of the militarist regimes that ruled Germany, Italy, and Japan during the Second World War echoed the behavior of the confederate states during the American Civil War. Each of these countries had been a member of the League of Nations (which Italy and Japan helped to found and which Germany joined in 1926), and had agreed to abide by its rules, including the rejection of aggressive warfare as an instrument of foreign policy. In the 1930s, each of these regimes opted to secede from the League of Nations in order to pursue a foreign policy based, not on the principles of federalism and international law, but on racial hierarchy and an ambitious program of territorial conquest.

The civil rights pioneer W. E. B. Du Bois saw a clear parallel between the Second World War and the American Civil War. Commenting on the UN Charter at the United Nations Conference on International Organization in San Francisco, Du Bois declared, "What was true of the United States in the past is true of world civilization today—we cannot exist half slave and half free" (Tuck 2012, p. 202). As in the Civil War, these regimes had to surrender unconditionally before the nation states they had ruled could return to the reconstructed system of federalism that was came to be called the United Nations. Since 1942, President Franklin Roosevelt had used this term to describe the growing list of nations from both hemispheres opposing the Axis powers, and by the spring of 1945 it became the name of the federalist organization that would replace the moribund League of Nations after the conclusion of the Second World War.

Of course, the federalist rhetoric that surrounded the creation of the UN was a far cry from the vision that Streit had advanced in the late 1930s. Having grown from a wartime alliance, the United Nations, in spite of its federalist veneer, would never pose a serious challenge to the most powerful nations in the wartime alliance, namely the United States, Great Britain, and the Soviet Union (now the Russian Federation). Along with France and China, each of these major powers has enjoyed a permanent place on the UN Security Council and can, provided its ambassador is present to vote, cast a veto against proposed UN Security Council resolution. Clarence Streit resumed his call for a political union of the world's democracies after 1945. In October of 1945, Streit participated in a conference in Dublin, New Hampshire to explore alternatives to the United Nations that would move the world closer to a viable world federation. Streit's view that a strong union of democracies must first be the nucleus of any future world government did not emerge as the majority view at the Dublin Conference, but he was joined in that position by Emery Reves and by Justice Owen J. Roberts (Baratta 2004b, pp. 543–544).

A New Hampshire Republican who had been appointed to the Supreme Court by Herbert Hoover, Justice Roberts had effectively opposed many of Franklin

Roosevelt's New Deal reforms because he saw them as an attempt by the Federal government to overstep its powers in relation to private enterprise. In 1937, however, Roberts cast a tie-breaking vote upholding a minimum wage law that had been passed by the state of Washington. This decision helped to avoid further confrontation between the White House and the Supreme Court and put an end to President Roosevelt's legislative proposal to add more justices to the bench of the Supreme Court. Justice Roberts has been cast as a pragmatist for breaking ranks with the laissez faire philosophy that had informed most of the court's earlier decisions during the Roosevelt era. According to this view, he voted as he did in 1937 in order to avoid the full blown constitutional crisis that might have arisen if the President had seen another piece of New Deal legislation struck down by the court's conservative majority.

It is certainly possible that such pragmatic considerations caused Roberts to break with the court's conservative faction in this case, but his broader record indicates that he was neither a conservative ideologue nor a malleable pragmatist. During World War Two, he refused to side with the court's majority when it supported the Presidents executive order requiring the internment of Americans of Japanese ancestry for the duration of the war. A maverick who proved to be part of no reliable voting block on the court, Roberts retired in 1945 and, because he had angered so many of his colleagues on the bench, he did not receive the customary letter of farewell and commendation from his colleagues. Two letters were drafted honoring Roberts' service, but neither proved to be a document that his supporters and detractors alike were willing to sign (Hall 2001, p. 298).

In his writings after 1945, Roberts argued that because governments derive their just powers from the consent of the people, the real source of sovereignty in any democratic polity must be the individual citizen. Concerned that attempts to establish new norms of international law in the wake of World War Two were being irrevocably distorted by concerns about "national sovereignty" Roberts took the remarkably radical position that this cherished concept is little more than a sacred cow, and that it is antithetical to both democratic governance and international law. Supporting Clarence Streit's idea for a federation of democracies to prevent future wars of aggression, Roberts wrote:

> National sovereignty is the enemy of international law. Its affirmation is the negation of law above the national level. The people must abandon the false doctrine of national sovereignty if they are to unite in world government of law. They must assert and exercise the sovereignty vested in each of them as human beings. So, and only so, can men of many nations form an enduring union of laws superior to their own national laws, for the protection and regulation of the interests they, as human beings, have in common. (Streit, Roberts, and Schmidt, 1950, pp. 12–13)

Because Roberts adhered to definition of political sovereignty that placed its source in the consent of the individual citizen, he could not be content with the structure of the United Nations as prescribed by the UN Charter.

The frustration of individuals such as Clarence Streit and Owen J. Roberts with the UN Charter and its deference to national sovereignty would be shared by other advocates of democracy and human rights for generations to come. For example,

Anne Marie Slaughter and G. John Ikenberry observed in 2006 that the UN Security Council, hamstrung by the veto power of its five permanent members, failed to respond to mounting transnational crises, and was reflective of a broader trend of "excessive bureaucracy, rigidity, and atrophy" that was coming to permeate the institution (Ikenberry and Slaughter 2006. p. 23). Drawn to a solution that echoed the thinking of Clarence Streit in the late 1930s, Slaughter and Ikenberry proposed a supranational organization that they called a "Concert of Democracies." Though this organization would not entail the full monetary, political, and military union of the world's democracies, it would entail a level of integration that is still unprecedented in order "to strengthen security cooperation among the world's liberal democracies and to provide a framework in which they can work together to effectively tackle common challenges" including climate change. (Ikenberry and Slaughter 2006, pp. 11, 25). Grand visions of uniting the world's democracies received much more publicity in the early 1950s, when Streit even graced the cover of *Time* magazine, but they still have a few influential adherents among foreign policy intellectuals today.

The attention given to Streit's ideas in the decade following World War Two reflected a widespread desire to prevent a new global conflagration that would be even more catastrophic than the last. As a reaction to the horrors of Hiroshima and Nagasaki, many scientists and public intellectuals called for more powerful institutions of global governance in order to prevent another great power war and to control the use of the atomic bomb. Albert Einstein, who together with the physicist Leo Szilard had called upon President Roosevelt to begin research on the construction of an atomic bomb in 1939, was one of the most influential advocates of creating some form of world government more powerful than the United Nations in order to prevent another war using nuclear weapons in the future. In 1947, Einstein sat down with the American journalist Raymond Swing to speculate about how such a world government might operate. Swing collected Einstein's remarks from this interview and published them in *The Atlantic* in an article called "Atomic War or Peace" in the autumn of that year.

Concerning the role that democratic institutions would play in the creation of such a government, Einstein took what he hoped would be a pragmatic approach:

> . . . [T]he representatives to supranational organization—assembly and council—must be elected by the people in each member country through a secret ballot. These representatives must represent the people rather than any government—which would enhance the pacific nature of the organization.
>
> To require that other democratic criteria be met is, I believe, inadvisable. Democratic institutions and standards are the result of historic developments to an extent not always appreciated in the lands which enjoy them. Setting arbitrary standards sharpens the ideological differences between the Western and Soviet systems. (Einstein 1950, pp. 197–198)

In light of the fact that the Marxist-Leninist regimes of the emerging Soviet bloc often used the language of democracy to describe their economic policies, Einstein attempted to avoid defining democracy beyond the practice of calling for representatives in a world government to by "elected by the people of each member country through a secret ballot." By limiting his employment of democratic institutions to what he hoped would be a functional minimum, Einstein was adhering to the same principle articulated by Owen J. Roberts, that sovereignty springs from the individual, and not from any national government. Einstein also hoped that future efforts to establish a democratic world government would avoid the minefield of ideology that was clouding East-West relations in the early years of the Cold War. He believed that a system of elected representatives would make it possible to work around such differences because he did not see ideology as the core of the problem:

> But it is not the ideological differences which now are pushing the world in the direction of war. Indeed, if all the Western nations were to adopt socialism, while maintaining their national sovereignty, it is quite likely that the conflict for power between East and West would continue. The passion expressed over the economic systems of the present seems to me quite irrational. Whether the economic life of America should be dominated by relatively few individuals, as it is, or these individuals should be controlled by the state, may be important, but it is not important enough to justify all the feelings that are stirred up over it. (Einstein, *Out of My Later Years*. New York: The Philosophical Library, 1950, p. 198.)

Two years later, after Stalin's regime had shown its resolute opposition to all forms of world federalism, Einstein argued that the democratic nations of the world should proceed to establish some form of transnational government without the Soviet Union. If such institutions were carefully crafted, Einstein believed, "Russia would cooperate once she realized that she was no longer able to prevent a world government anyhow" (Isaacson 2007, p. 498). In some sense, the history of relations between Moscow and Washington since the collapse of the Soviet Union has given credence to Einstein's contention that ideology was not the core of the problem. Even though capitalist institutions thrive in Russia, many of the old conflicts between the United States and Russia over places such as the former Yugoslavia, Syria, and Ukraine have remained quite serious, suggesting that the broad contours of geopolitical tensions often remain the same in spite of changes in economic practices or ideology. On the other hand, Einstein's presumption that free and fair elections could be conducted without a number of other democratic institutions in place seems naïve. Unless a nation has such safeguards as a free press, an independent judiciary, and a competitive multi-party system, it is not likely to sponsor the free and fair elections via "secret ballot" that Einstein saw as the necessary foundation for the establishment of a viable global democracy.

Among scientists and intellectuals, Einstein was far from alone in his desire for the creation of a cosmopolitan democracy. In 1946, a concerned group of leading physicists and public intellectuals issued a call for world government in a short collection of essays entitled *One World or None*. Written before the chasm between the United States and the Soviet Union would widen over the crises of the late 1940s, the essays in *One World or None* called for the establishment of institutions of global governance that would be sufficient not only to prevent another world war, but also to head off a global arms race in the production of nuclear weapons. In his contribution to the volume, Einstein argued that, in order to possess lasting legitimacy, such international institutions had to be democratic in character. The seminal journalist Walter Lippmann, who also contributed an essay to the volume, advanced the argument that any global authority capable of ensuring justice and preventing war would have to derive its legitimacy from its relationship to individual citizens as opposed to sovereign nation states. Lippmann pointed to the failure of the League of Nations in the 1930s as a clear example of shortcomings of collective security arrangements when they are conceived exclusively as an arrangement between sovereign nation states. Pointing to the success of the Nuremburg Trials, Lippmann advanced the idea that world law had already emerged upon the scene, and that the precedent set by the international proceedings against Nazi war criminals had established a general principle upon which an entirely new set of international practices could be tried and expanded upon.

The writer E. B. White was another prominent advocate of world federalism in the 1940s. As the lead writer behind *The New Yorker*'s "Talk of the Town" column during this period, White began advocating world federalism in 1943, citing Emery Reves as his primary inspiration (White 1946, p. 4). As he covered the development of the United Nations in New York City in 1946, White made his own case that an organization of sovereign nations was an insufficient answer to the challenges posed by the age of the atom bomb. Reviewing a discussion of spheres of influence and the newly formed UN Security Council, White observed wryly, "There is only one sphere of influence today, and we live on it" (White 1946, p. 170). Like Reves, White argued that democracy would not survive if human beings did not create democratic institutions beyond the boundaries of the nation state.

White's conception of democracy involved free and fair elections, a free press, and an independent judiciary, but it also entailed a host of intangibles that he described in the summer of 1944: "Democracy is the recurrent suspicion that more than half of the people are right more than half of the time. It is the feeling of privacy in the voting booths, the feeling of communion in the libraries, the feeling of vitality everywhere." Alluding to the suspense of a good baseball game, or perhaps to the genius of Beethoven, he concluded: "Democracy is the score at the beginning of the ninth" (White 1946, p. 31). As he advocated the expansion of democracy on a global scale, E. B. White was certainly aware that many would dismiss such proposals as unrealistic. In his last essay on the subject, he assessed the question of whether the world was ready for the creation of a constitutional government "on a planetary scale." He concluded that, "it will never be ready. The test is whether people will chance it anyway. .." (White 1946, p. 185).

In White's time, as now, a major stumbling block to creating a supranational democracy has always been a sense of national pride, often justified by appeals to military glory. Like other forms of communal identity, national pride is likely to be a powerful force in human affairs for a long time to come. As Yascha Mounk put it, "Nationalism is like a half-wild, half domesticated animal. As long as it remains under our control. .. it can genuinely enrich our lives" but when "it breaks free of the constraints we put on it. .. it can be deadly" (2018. p. 215). Some historians have observed, however, that the citizens of a nation can also take pride in the efforts of their country to establish the rule of law among nations. In the case of the United States, for example, Michael Bess has observed, "Those who argue, today, that supporting the United Nations is 'un-American' need to sit down for a few hours with a good history textbook" because multilateralism and the construction of international institutions constitutes "a long tradition of American foreign policy, a tradition reflecting the convictions held by some of the most eminent Republican and Democratic statesmen of the past 120 years" (2006. p. 283). Just as respect for one's own family is not mutually exclusive with respect for one's community, respect for one's own nation is not mutually exclusive with respect for the larger community beyond that nation. In fact, each depends vitally upon the other.

As noted earlier, another alternative to the United Nations emerged at the University of Chicago, soon after the bombings of Hiroshima and Nagasaki. Deeply disturbed by the power of the new weapon that his university had helped to create, University of Chicago president Robert Maynard Hutchins placed his faith in the concept of World Federalism. At the suggestion of the Italian antifascist writer and scholar Giuseppe Antonio Borgese, Hutchins convened a Committee to Draft a World Constitution. The proceedings and debates that emerged from this group were chronicled in a journal called *Common Cause*. The democratic constitution that this committee drafted was published in 1948. Before leading the Committee to Draft a World Constitutions, Robert Maynard Hutchins had transformed the University of Chicago with a curriculum that emphasized canonical texts that he deemed to be of perennial value to all. Although his approach to higher education was far more conservative than that of other twentieth century intellectuals such as John Dewey, Hutchins described his own educational philosophy as grounded in a fundamentally democratic and egalitarian faith in the ability of all people to advance through education:

> The business of saying... that people are not capable of achieving a good education is too strongly reminiscent of the opposition of every extension of democracy. This opposition has always rested on the allegation that the people were incapable of exercising the power they demanded. Always the historic statement has been verified: you cannot expect the slave to show the virtues of the free man unless you first set him free. When the slave has been set free, he has, in the passage of time, become indistinguishable from those who have always been free... There appears to be an innate human tendency to underestimate the capacity of those who do not belong to "our" group. Those who do not share our background cannot have our ability. Foreigners, people who are in a different economic status, and the young seem invariably to be regarded as intellectually backward. . . (*Hutchins* 1954, p. 45)

In linking the extension of education to the extension of democracy, Hutchins affirmed the same faith that had been the basis for his strategy to establish international institutions that would assure peace in the age of nuclear weapons. Within a week of the bombing of Hiroshima, Hutchins made an appeal on an NBC radio broadcast for the creation of a world government to prevent the destruction of civilization. In a pamphlet which he published before the end of 1945 entitled *The Atomic Bomb vs. Civilization*, Hutchins argued that the United States should honor its commitment to the United Nations Charter, but that much stronger institutions of global governance would be necessary to prevent a nuclear arms race and nuclear war in the future. Inspired by the activism of Borgese, Hutchins formed a committee that included Mortimer Adler, Harold A. Innis, and other prominent academics to set about producing a draft constitution for a democratic and effective world government. The committee released its constitution before the end of the decade, but by 1949 Cold War tensions had rendered ideas about world government politically suspect within the United States.

When the Hutchins Committee issued its *Preliminary Draft of a World Constitution*, nationalist sentiments were beginning a new period of resurgence that would continue for decades to come. The deepening of Cold War divisions, exacerbated by the Berlin crisis of 1948–49, strengthened nationalist feeling on both sides of the Iron Curtain. At the same time, the rising tide of anti-colonialism after World War Two provided fertile ground for new nationalist movements across Asia and Africa. This changing landscape was reflected in the response to the committee's work, even among individuals who were inclined to be sympathetic to the project. Soon after the draft was issued, Prime Minister Nehru of India and Wellington Koo, the Chinese ambassador to the United States, joined Hutchins in a Round-Table of the Air discussion of "The Problem of World Government." In his book *The Discovery of India*, Nehru had argued cogently that the logic of empire and "world conquest" was no longer viable and must be replaced by an ethos of "world association." He critiqued the realist concept of national interest as outdated in an age when "The interests and activities of states overflow their boundaries and are world-wide" (Nehru 1946, p. 602). However, by 1949, the landscape for Asia and the world was considerably less hospitable to broad schemes for global governance, however inspired and intelligent they might seem. Ambassador Wellington Koo warned that Nationalist China, which was torn by civil war and was on the verge of losing the entire mainland to the Chinese Communist Party was in no position to consider the ideas formulated in Chicago regarding a world constitution (Mayer 1992, p. 279). For his part, Nehru admitted that India's triple preoccupation with its economic distress, its division into two separate nations, and, above all, its newly won national independence from Great Britain had not created a fertile ground for the reception of world federalism in India. Nonetheless, Nehru would remain committed to the ideals of the "One World" movement well into the 1950s, as exemplified by his 1956 address to the UN General Assembly, "Towards a World Community" delivered in response to the crises in Hungary and the Suez Canal that year (Bhagavan 2012, p. 157).

For public opinion in the United States, however, the publication of *One World or None* right after World War Two had probably been the high-water mark of the world federalist movement. In August of 1949 the Soviet Union tested its own nuclear bomb in Siberia, thus ending the American monopoly on this super weapon, opening a new kind of arms race, and putting an end to any high-minded discussion about international control of this dangerous new technology. By October of 1949, the forces of the Mao Zedong would have the whole of mainland China under their effective control, while the American-backed Guomindong government, led by the military strongman Chiang Kaishek, would retreat to the island of Taiwan. The double jolt of these events pushed all further discussion of the ideas discussed in the pages of *One World or None* far into the margins of American political discourse. In the public sphere, political opportunists such as Joseph McCarthy exploited a widespread fear of domestic communist subversion, tarring the reputation of two of the contributors to *One World or None*, Robert Oppenheimer and E.U. Condon, in the process.

Within the Truman administration, the dramatic events that had closed the curtain on the 1940s also shifted its policy initiatives sharply away from anything that resembled the spirit of internationalism that had reached its peak in 1946. In June of that year, Truman had proposed the creation of international institutions to control nuclear technology under the auspices of the Baruch Plan. Although Truman most likely expected Stalin to reject the plan, the fact his administration offered it at all indicated an awareness that it was at least wise to lend lip service to the idea of international control of the atomic bomb. By the spring of 1950, Truman was pushing ahead, on the urgent advice of his National Security Council, with the creation of the hydrogen bomb and the expansion of covert operations by the Central Intelligence Agency (NSC-68. April 7, 1950). In the summer of 1950, when the U.S. dispatched troops to the Korean peninsula to counter an invasion of South Korean by North Korean forces, the President obtained approval from the UN Security Council, but only because the Soviet representative was not present and therefore unable to implement the veto power that the USSR held as a permanent member. Although it had at least a nominal imprimatur from the UN Security Council, the Korean War accelerated a resurgence of nationalism. The conflict heightened Cold War tensions far beyond East Asia, and dealt a serious blow to the idea of world federalism (Baratta 2004b, p. 486). Even in the midst of this crisis, however, Truman still talked wistfully about the dream of world government sketched in Tennyson's "Locksley Hall." Riding in his limousine after an early morning walk with the author John Hersey in 1951, Truman took the folded piece of paper out of his wallet and let Hersey read it. "Notice the part about universal law," said the president. "We're going to have that some day. I guess that's what I've really been working for ever since I first put that poetry in my pocket" (Strobe Talbott 2009, p. 210).

## Bibliography

Appiah, Kwame Anthony. 2006. *Cosmopolitanism: Ethics in a world of strangers*. New York: W.W. Norton & Co.

Atkinson, David C. 2016. *The burden of white supremacy: Containing Asian migration in the British empire and the United States*. Chapel Hill: The University of North Carolina Press.

Bailey, Ronald. 1994. *Ecoscam: The false prophets of ecological apocalypse*. New York: St. Martin's Press.

Baratta, Joseph Preston. 2004a. *The politics of world federation*. Vol. 1. Westport: Praeger.

———. 2004b. *The politics of world federation*. Vol. 2. Westport: Praeger.

Berlin, Isaiah, and Henry Hardy, eds. 2013. *The crooked timber of humanity: Chapters in the history of ideas*. Second ed. Princeton: Princeton University Press.

Bhagavan, Manu. 2012. *The peacemakers: India and the quest for one world*. New Delhi: Harper Collins & The India Today Group.

Chan, Wing-tsit. 1963. *A source book in Chinese philosophy*. Princeton: Princeton University Press.

de Condorcet, M. 1795. *Sketch for a historical picture of the progress of the human mind*. Trans. June Barraclough. New York: Noonday Press, 1955.
De Vitoria, Francisco. 1991. In *Political writings*, ed. Anthony Pagden and Jeremy Lawrence. Cambridge: Cambridge University Press.
Eaton, Casindania P. , et al. n.d. "Peace by law our one hope" letter to the *New York Times* Oct 10, 1945. p. 20.
Einstein. 1950. *Out of my later years*. New York: The Philosophical Library, Inc. 1950.
Eisenhower, Dwight. 1953. "*The chance for peace*" April 16th, 1952. Eisenhower Presidential Library. https://www.eisenhower.archives.gov/all_about_ike/speeches/chance_for_peace.pdf
Grotius, Hugo. 2009. *Hugo Grotius Mare Liberum 1609–2009: Original Latin Text and English Translation*. Leiden: Bril.
Hall, Timothy L. 2001. *Supreme court justices: A biographical dictionary*. New York: Facts on File.
Hamilton, J.R. 1999. *Plutarch: Alexander*. London: Duckworth Publishing.
Hesiod. 2017. *The poems of Hesiod: Theogony, works and days, and the shield of Herakles* Trans. Barry B. Powell Oakland: University of California Press.
Holden, Barry, ed. 2000. *Global democracy: Key debates*. London: Routledge.
Hutchins, Robert Maynard. 1954. *Great books: The foundation of a liberal education*, 45. New York: Simon & Schuster.
Isaacson, Walter. 2007. *Einstein: His life and universe*. New York: Simon and Schuster.
Kant, Immanuel. 1991. In *Kant: Political writings*, ed. Hans Siegbert Reiss. Cambridge: Cambridge University Press.
Kant, I., P. Kleingeld, J. Waldron, M.W. Doyle, and A.W. Wood. 2006. *Toward perpetual peace and other writings on politics, peace, and history*. New Haven: Yale University Press.
Kaplan, Temma. 2015. *Democracy: A world history*. Oxford: Oxford University Press.
Mayer, Milton. 1992. *Robert Maynard Hutchins: A memoir*. Oakland: University of California Press.
Mazower, Mark. 2009. *No enchanted palace: The end of empire and the ideological origins of the United Nations*. Princeton: Princeton University Press.
———. 2012. *Governing the world: The history of an idea*. New York: The Penguin Press.
Monnet, Jean. Richard Mayne, trans. 1978. Memoirs. New York: Doubleday & Company.
Nehru, Jawaharlal. 1946 Reprint 2004. *The discovery of India*. Penguin Books.
Orwell, George. [1939] 1968 "Not counting niggers" in *Collected essays, journalism, letters George Orwell: Age like this,* 1920–1940. New York: Harcourt, Brace, and Company.
———. [1945] 1968. "You and the atomic bomb" in Sonia Orwell and Ian Angus, eds. In *Front of your nose,* 1945–1950. London: Secker & Warburg.
Owen, John M. 2010. *Religion, the enlightenment, and the new global order*. New York: Columbia University Press.
Reves, Emery. 1945. *The anatomy of peace*. New York: Harper and Brothers.
Ricks, Thomas. 2017. *Churchill and Orwell: The fight for freedom*. New York: Penguin Press.
Rosenboim, Or. 2017. *The emergence of globalism: Visions of world order in Britain and the United States, 1939–1950*. Princeton: Princeton University Press.
Schweickart, Russell L. 2011. *Personal interview via email*. 19 Dec 2011.
Scott, James Brown. 2000. *Francisco de Vittoria and His Law of Nations*. Union: The Lawbook Exchange, Ltd.
Stella, Tiziana. 2017. *Personal interview via email*. 20 July 2017.
Streit, Clarence K. 1939. *Union now: A proposal for a federal union of the democracies of the North Atlantic*. London: Jonathan Cape.
Talbott, Strobe. 2009. *The great experiment: The story of ancient empires, modern states, and the quest for a global nation*. New York: Simon & Schuster.
Tennyson, Alfred.1842. "Locksley hall" *poems of Tennyson*.
Tuck, Steven. 2012. *Fog of war: The second world war and the civil rights movement*. Oxford: Oxford University Press.
White, E.B. 1946. *The wild flag*. New York: Houghton Mifflin Company.

# Chapter 8
# Climate Change and the Future of Democracy

*"We have frequently printed the word Democracy. . . .It is a great word, whose history, I suppose, remains unwritten, because that history has yet to be enacted."*

Walt Whitman

**Abstract** There are three broad paths to establishing supranational democracy in order to address the transnational challenge of climate change. The first and most direct path would involve bypassing national governments altogether and organizing municipalities on a global scale. The second and most cautious path would involve creating an integrated federation of established democracies that could grow over time. The third and most ambitious path would involve democratizing already existing global institutions such as the United Nations. Each of these paths has benefits and drawbacks, but the middle path of pursuing greater political integration among established democracies is probably the most viable.

**Keywords** Parliament of mayors · UN parliamentary assembly · *Pax democratica*

The challenges of nuclear proliferation and environmental crises since 1945 has caused so many to lament that our technical progress has outstripped our moral evolution that the observation has itself has become a cliché. It may also be an oversimplification. In *The Better Angels of Our Nature*, the psychologist and linguist Steven Pinker makes the case that human beings have become steadily less violent since the advent of agriculture and state building, and that they have become precipitously less violent since the spread of literacy and the Enlightenment values of rationalism and tolerance over the past three centuries. Pinker bases his argument on a Himalayan mass of statistics, the sum of which point to the conclusion that any person born today is far less likely to die a violent death at the hands of another person than was the case in the past. This was even the case in the twentieth century, in spite of such abysmal developments as World War One, World War Two, and the



R. S. Deese, *Climate Change and the Future of Democracy*, Environmental Challenges and Solutions 5, https://doi.org/10.1007/978-3-319-98307-3_8

Holocaust. Pinker attempts to explain this trend by exploring a variety of factors, but the greatest of these concerns the changing ethos that has informed human interaction over the course of millennia. Affirming the reasoning of Thomas Hobbes, Pinker argues that people who are living in a society in which the state's power to enforce law is weak or nonexistent must embrace an ethos of revenge in order protect themselves. This ethos of revenge, in order to remain credible, must be affirmed by a frequent recourse to acts of violence, even when those acts of violence might not serve one's rational self-interest.

In other words, if a person wants to protect his vital interests in a lawless society, he needs to create a reputation for retaliation, and, in order to maintain the ironclad credibility of that reputation, he must exercise violence against those who merely attack his reputation, long before they present any tangible danger to himself, his kin, or his material interests. The logic of this kind of "deterrence" gives rise, as Pinker describes it, to a kind of honor culture in which an ethos of hair-trigger retaliation becomes commonplace. Conversely, in a society where the state possesses a clear monopoly on the legitimate use of violence, an ethos of civility and nonviolence among its citizens takes root and begins to flourish. And, as a secure society gradually becomes more educated and empathic, two historical trends for which Pinker gives a great deal of credit to the Enlightenment and the proliferation of the modern novel, people become even less likely to resort to violence to protect their honor or to settle disputes.

This gradual reduction of violence has been essential to the advance of scientific knowledge and democratic governance, but it cannot be taken for granted. Arguing that the values of science and democracy are inextricably linked, the physicist and historian of science Timothy Ferris sees a direct correlation between the rejection of science and rationality and the rejection of democracy. Identifying the works of Jean Jacques Rousseau as the fountainhead of Romanticism, Ferris argues that Rousseau's rejection of rationalism and individual rights were the direct inspiration for the tyrannical policies of Robespierre and the outburst of paranoia and bloodletting unleashed by the Reign of Terror during the French Revolution (Ferris 2011). The totalitarian regimes of the twentieth century, Ferris observes, followed a similar pattern of rejecting the Enlightenment principle of rational discourse and promoting new forms of fanatical devotion to group identities and to the cults of personality developed by such tyrants of the left and right as Stalin, Mussolini, Hitler, and Mao.

Steven Pinker identifies a common denominator of such anti-democratic movements as the "counter-Enlightenment idea that people are expendable cells in a superorganism. .. and that the supreme good is the glory of the collective rather than the wellbeing of the people who make it up" (2018, p. 31). Although such dedication to the collective is usually associated with religious or political extremism, he warns that the ubiquitous culture of nationalism has the potential to devolve from an admirable sense of responsibility and dedication to one's community into a dangerous devotion to a sterile abstraction. Pinker elaborates this distinction as follows: "When a 'nation' is conceived as a social contract among people sharing a territory. .. it is an essential means for advancing its members' flourishing"; conversely, when the entity of the nation-state becomes an object of increasingly irrational idolatry, the individual citizen is soon "forced to make the supreme sacrifice for the benefit of a charismatic leader, a square of cloth, or colors on a map" (2018, p. 31). Just as Emery Reves decried the irrational potential of nationalism in the twentieth century, Pinker warns that the worship of the nation state, along with religious extremism, "continue[s]. . . to affect the fate of billions of people" in the twenty-first century (2018, p. 31).

In spite of such recurring setbacks, there is clear evidence that the values of rational discourse, individual rights, and democratic accountability have shown remarkable resilience over the past three centuries. In light of the decline in violence that Pinker has linked to the slow but steady proliferation of Enlightenment values, it is at least reasonable to wonder whether the rise of international law could reform the psychology and behavior of nation states in a similar manner. As long as nations exist in a Hobbesian "state of nature" in their relations with each other, it would seem likely that they would frequently resort to violence, not only to protect their borders and their tangible interests, but also to protect their reputation on the international stage. If the Hobbesian paradigm holds true for nations, the only way to prevent a continued war of "all against all" among nations would be to create a new Leviathan in the form of a global superstate that would hold all of the individual states in awe of its power and compel them to obey a clear code of international law.

The scholarship of Pinker and Ferris points to a virtuous cycle: The values of science and democracy benefit from peace, and, if they are preserved and expanded, they help to keep the peace. The power of this cycle to safeguard our common interests may prove to be more significant than the stabilizing force of any empire or superstate. In the seven decades that have followed the advent of atomic weapons, a global Leviathan that would fully establish the rule of law among nations has not emerged, and yet neither has another great power war that comes anywhere close to the scale of the first or second world wars. The United Nations, which some had openly hoped would become this new Leviathan in the late 1940s, has remained largely powerless to restrain the behavior of any of the major powers that comprise its security council. In spite of the superpower pretensions of both the United States and the Soviet Union, neither Washington nor Moscow ever possessed enough power after 1945 to impose upon the world anything equal to the Pax Britannica that prevented the outbreak of a major war among the great powers for almost a century between the fall of Napoleon and the outbreak of the First World War. The mutual annihilation guaranteed by nuclear weapons is at least part of the explanation for why the great powers of the world have not engaged in direct warfare with each other over the past several decades, but the growing influence of public opinion in international affairs also deserves consideration.

The story of Alfred Nobel and Bertha von Suttner illustrates this point. The idea that a super-weapon would render war obsolete was commonplace even in the late nineteenth century, when Alfred Nobel speculated that his invention of TNT might have that effect. In 1891, Nobel had said to the writer and pacifist activist Bertha von Suttner, "Perhaps my factories will put an end to war sooner than your congresses: on the day that two army corps can mutually annihilate each other in a second, all civilized nations will surely recoil with horror and disband their troops" (Tägil 1998). Of course, the First World War would prove that the chemical high explosives developed by Nobel did not prevent war, but only made it many times more destructive than it had ever been in the past. Nobel's long acquaintance and correspondence with Berth von Suttner, however, would heighten his desire to prove that his fortune, which owed so much to the sale of war material, could be put to good use in serving the cause of peace. Nobel had known Suttner since the 1870s, when she had served for a time as his secretary in Paris, and he was deeply impressed by her dedication to the cause of peace (Blom 2008, p. 190). When he drafted his will, and set aside funds for the creation of the Nobel Peace Prize, he was eager to tell von Suttner of his decision. She responded to him promptly and claimed a share of the credit for influencing his decision, "Whether I am around then or not does not matter; what we have given, you and I, is going to live on" (Tägil 1998).

The drama of Nobel's gradual transition from military industrialist to the world's best-known philanthropist for the cause of peace around the world is dramatic but hardly implausible when we consider his very human desire to be thought well of, both by the charismatic young peace activist who influenced him, and by people around the world. If we frame this concern for one's own reputation in personal terms, we might call it vanity, or at the very least the possession of a sense of shame. If we were to frame it in less personal terms, so that we might observes its importance to conglomerations of individuals as well to individuals themselves, we might call it political capital. Unlike economic capital, political capital may not be measurable in discrete units of Dollars, Euros, or Bitcoin, but it is no less real for that. On an individual scale, we sense the value of political capital implicitly. On the scale of corporations, societies, and nation states, the value of political capital is less widely recognized, but its salience has grown considerably with the steady globalization of commerce and communication. In the realm of interpersonal relationships, it is probably concern for one's own political capital, and not just the fear of law enforcement, the motivates most people to treat one another with at least a passable level of decency. For example, if the manager of a small group of employees wants to implement a change of practice in an office environment she would have several options for effecting that change. She could call a meeting in which she browbeats her staff and threatens them with dismissal if they do not immediately implement the change. This would likely mean that the change would in fact be implemented rapidly, but with the loss of so much political capital as to make the cost of such a high-handed approach too great. The best employees at the office might find that work environment intolerable and leave for opportunities elsewhere, while those employees compelled by circumstances to remain would manifest their discontent in a variety of ways that would be detrimental not only to the specific goals of the unpopular manager, but to the larger needs and goals of the institution itself. On a global scale, the loss of political capital is no less of a problem. When great powers have attempted to pursue their foreign policy goals without "a decent respect to the opinions of mankind," they inevitably suffer a loss of political capital that will prove to be more detrimental than their leaders had ever imagined. This happened to Britain and France as a result of their precipitation of the Suez Canal crisis in 1956, and it happened on a much larger scale to the United States and Soviet Union as they waged long unilateral wars in Vietnam and Afghanistan.

The fact that nation states, however much they may view their sovereignty as sacred and inviolate, are likely to be called to account for their actions in a global court of public opinion is one of the things that has rendered the pretense of empire increasingly ridiculous in the modern age, especially as the diffusion of real time information has become harder for any centralized system of authority to control. Empiricism has proven to be the enemy of empire. The rise of empiricism, facilitated by horizontal modes of communication, has rapidly dissolved the stability of empires, which have always required the vertical imposition of uniform thought and behavior by a single center of power and authority.

A recent example of the contrast between the lowly practice of empiricism and the arrogant practice of empire-building can be seen in this widely-quoted exchange between the journalist and author Ron Suskind and a highly-placed advisor to President George W. Bush concerning the 2003 U.S. invasion of Iraq. In that exchange, the unnamed Bush advisor (who has since been widely identified as Karl Rove), presents his down-home version of a Foucauldian discourse on the relationship between knowledge and power. Dismissing the work of those who try to take all the facts into consideration in order to understand situations and formulate policy, this White House sage argues that empires do not need to take account of facts, because they are in a position to create their own reality:

> The aide said that guys like me were "in what we call the reality-based community," which he defined as people who "believe that solutions emerge from your judicious study of discernible reality." I nodded and murmured something about enlightenment principles and empiricism. He cut me off. "That's not the way the world really works anymore," he continued. "We're an empire now, and when we act, we create our own reality. And while you're studying that reality—judiciously, as you will—we'll act again, creating other new realities, which you can study too, and that's how things will sort out. We're history's actors . . . and you, all of you, will be left to just study what we do" (Suskind 2004, p. 17).

Of course, as John Adams once said, "Facts are stubborn things." In the years that have passed since the verbal exchange that Suskind describes above, the empiricists have won the consolation prize that they usually win when the empire builders choose to ignore them, the general acknowledgement that they were right all along, and that a terrible series of disasters could have been avoided if only the empire-builders had listened to them. The privilege of saying "I told you so" is virtually worthless, but the right to say "Listen to me next time if you don't want to cause the same disaster twice" is indispensable.

In the realm of science, the ethos of empiricism demands rigorous peer review; in the realm of politics, it demands a free press and competitive elections. There are two broad questions about government facing the human race in the Anthropocene. The first is, can we govern nature? The second is, can we govern ourselves? The most probable answer to the first of these questions, in spite of our advances in technology, is no. The only honest answer to the second question is maybe. While the threat posed by climate change does not entail the sudden and total destruction of nuclear war, it does promise to impose incredible stress on the liberal democracies of the world through an increased frequency of powerful and destructive storms, crop failures, mass migrations, and unprecedented droughts. The United States has already proven that its electoral system is vulnerable to these challenges. Over the past two decades, a discernible feedback loop has emerged between disruptions in the U.S. electoral system, its climate policies, and severe weather events related to climate change. Many factors cause extreme weather events, but the most significant factor that we can address is the role played by greenhouse gases. Many factors cause election outcomes in which the intent of the voters is not faithfully recorded, but the most significant factors that we can address is a lack of uniform balloting standards, compounded by state and local legislation that "collectively reduce electoral access among the socially marginalized" (Bentele and O'Brien, 2013). Unfortunately, the health of democracy around the world is gravely threatened by the declining health of democracy in America. Since the 2000 U.S. Presidential election, federal and state governments in the United States have made only piecemeal efforts to address either of these problems. Because the climate *in which* people live and work and the system of government *under which* they live and work interact in myriad ways, an increase of disruptive events in one system will likely lead to an increase of disruptive events in the other, creating a positive feedback loop that threatens a precipitous destabilization of both the world's climate and America's struggle to maintain a democratic system of governance. On one side of this feedback loop, the increase in major storms, destructive fires, and crop failures is likely to weaken democratic institutions and hasten the advance of corporate power and authoritarian governance on the state and federal level. On the other side, distortions in the American balloting process as witnessed in states such as Florida, Missouri, New Mexico, and Ohio since the beginning of the 2000 Presidential election have already weakened efforts to address the problem of climate change on a state, national, and international scale.

In the case of the 2000 U.S. Presidential election, for example, the electoral victory of George W. Bush, based on a seriously flawed balloting process in Florida and the gerrymandered structure of the Electoral College, led to an outcome that did not reflect the popular vote (Lynch 2001, pp. 417–419). The new administration implemented an immediate U.S. abrogation of the Kyoto Protocol and a dramatic reduction in federal efforts to address climate change (or to even acknowledge basic climate science), across the United States. The rapid U.S. withdrawal from the Kyoto Protocol paved the way for other major powers, such as Canada, Russia, and Japan to withdraw from the treaty in 2011. In other words, a local factor of a flawed balloting process in Florida helped to kill the Kyoto protocol in the United States, and this came to have a major impact on the course of climate policy across the globe.

Just as "the perfect storm" of the flawed Florida balloting process contributed to changes in global climate policy, it is also possible to see how a changing climate has produced weather events that can affect the outcome of elections. In the twenty-first century, disruptive weather events such as Hurricane Katrina in 2005and Super Storm Sandy in 2012 have shown that climate change can influence electoral outcomes. The federal government's widely criticized response to Katrina hurt the administration of George W. Bush and the electoral fortunes of his political party in the 2006 and 2008 elections. Conversely, the widely applauded federal response to the aftermath of Super Storm Sandy in 2012 seemed the help the administration of Barack Obama and the electoral fortunes of his party in the 2012 election. In the 2016 election, the vulnerabilities of the U.S. electoral system were even more conspicuous than they had been in 2000. The hacking of both major political parties by a foreign government and extensive misinformation campaigns, especially on social media platforms, marred the campaign itself while gerrymandering, voter suppression, and the antidemocratic institution of the Electoral College distorted the outcome of the balloting. For the second time in less than twenty years, the winner of the Electoral College actually lost the popular vote, this time by a margin of 2%, or about three million votes. And, as in 2000, the chaos engendered by the American electoral system immediately spread to climate policy. Where George W. Bush had used his Electoral College victory to abrogate all U.S. commitments to the Kyoto Accords of 1997, Donald J. Trump used his unexpected victory in the Electoral College to unilaterally nullify U.S. commitments to the Paris Accord of 2015.

The kind of chaos that the United States has witnessed in two of its last four elections is a harbinger of more to come. The severe disruptions caused by climate change are likely to create conflicts and refugee crises that will be far beyond the capacity of individual nation states, including the United States, to address. The strengthening of existing transnational institutions or the creation of new ones will be a likely outcome of the coming climate crisis. Given the integrative power of economic globalization, these institutions could gradually coalesce into a system of governance that, whether it calls itself a "world state" or not, will be an economic and political Leviathan of unprecedented scope and power. That Leviathan may prove capable of protecting the global networks of commerce and communication on which industrial civilization now depends, but it will not respect the democratic values of individual liberty, transparency, and accountability unless informed and committed people demand that it does so from the start.

Barring a complete collapse of civilization, some form of global government is coming to the planet Earth, probably within the next century. As Emery Reves observed in 1945, the logic of industrialism is drawing the world into a single integrated system: "Industrialism tends to embrace the whole globe in its sphere of activity." This was not the result of any lofty idealism on the part of industrialists, but due to the simple fact that "Modern industrial production needs raw materials from every corner of the earth, and seeks markets in every corner of the world" (Reves 1945, p. 143). In direct opposition to the process, Reves observed, "Nationalism. . . tends to divide the world. .. and segregate the human race into smaller and smaller independent groups" (1945, p. 143). As a man who had witnessed the dissolution of the Austro-Hungarian Empire as a youth and then lost his mother to a campaign of mass-murder waged by Hungarian fascists, Reves was speaking from personal experience (Gilbert 1997, pp. 15–17). Surveying the destruction of two world wars, Reves concluded that the intensifying conflict between the integrative logic of industrialism and the divisive passions of nationalism was the root of the problem: "It is the collision between our political life and economic and technological life that is the cause of the twentieth century crisis with which we have been struggling since 1914, as helpless as guinea pigs" (1945, pp. 143–144). These words could also apply to the twenty-first century crises of climate change, economic dislocation, and resurgent nationalism.

Supporters of democracy cannot protect democratic values by clinging to the model of national sovereignty that conceives of the autonomous nation state as the best haven for protecting democracy. As global institutions continue to coalesce and grow more powerful, the only viable strategy for the protection of democracy is to demand that those institutions are subjected to democratic accountability on a global scale. In the late 1940s the spirit of international cooperation inspired by the horrors of World War Two helped to create a phalanx of global institutions that have not only survived but, especially in the case of the IMF and World Bank, have grown in power and influence. However, none of these institutions has become more democratic in any measurable sense. Meanwhile, the institution of the liberal nation state, which provided an essential shelter for the evolution of democracy in the nineteenth and early twentieth centuries has become increasingly inimical to the democratic values as the Cold War machinery of "national security" has grown even more intrusive in an age of international terrorism. For democracy to survive, it must not only struggle to preserve democratic practices on the level of the nation state, but must also work to build new democratic institutions on a global scale. As noted in Chapter Seven, the movement that sought to assure the survival of our species in the face of nuclear war declared, "One World, or None." The emerging movement that seeks to protect democracy in the face of climate change and its attendant economic and political crises is driven by the apprehension of a truth that could be expressed in similar terms: "One Democracy, or None."

When the end of the Cold War came in the late eighties and early nineties, there was indeed a surge of global institution building, but it reflected different aspirations than those that had animated the world federalist movement of the late 1940s. The creation of the World Trade Organization in the last decade of the twentieth century was emblematic of the faith in the power of market forces not only to create wealth and technological innovation, but also to integrate the entire world through the expansion of commerce. The "neo" in "neoliberalism" reflected the fact that economists such as Milton Friedman and leaders such as Margaret Thatcher and Ronald Reagan had revived this faith in the power of unregulated markets in Great Britain and the United States, ending decades of Keynesian consensus. However, the most significant embrace of unbridled capitalism took place outside of the west, as countries as diverse as China, Russia, Mexico, and India also embraced the deregulation of markets and the privatization of economic enterprises that had previously been controlled entirely by the state. The widespread embrace of neoliberal economic policies combined with the rapid growth of new communication technologies such as the Internet and mobile phones lend credence to the notion that commerce and communication alone might integrate the world into a single community where the earnest efforts of international organizations such as the United Nations had failed. Prior to the First World War, Norman Angell had argued that the accelerated integration of the economies of Europe had made war absurd and unprofitable. In the summer of 1914, Angell's reasoning had been ignored by the leading statesmen of Europe but at the end of the twentieth century his ideas seemed as though they might be vindicated not only in Europe but also on a global scale. As the official integration of European economies signaled a triumph of commerce and cooperation over nationalism and war, the creation of the World Trade Organization in January of 1995 pointed to the promise that a new global network of commercial partners could also create lasting institutions that would help to secure not only prosperity but also peace on a global scale.

In 1997, as the centrist Democrat Bill Clinton began his second term as president and former Soviet leader Mikhail Gorbachev set a new precedent by playing himself in a television advertisement for Pizza Hut, the neoliberal vision of globalization seemed to have reached its zenith. Embracing many of the neoliberal policies of his Republican predecessors Reagan and Bush, Clinton had overseen the implementation of the North American Free Trade Agreement over the objections of organized labor, deregulated the pharmaceutical industry to speed new drugs to market and allow their advertisement on television, and would soon sign legislation that would leave the banking industry with less government oversight than at any time since the 1920s. Though widely seen as a financially desperate move on his part, Mikhail Gorbachev's Pizza Hut advertisement could not help but have ideological overtones for its Western audience (due to Gorbachev's unpopularity in Russia, the ad was only slated for broadcast outside of his home country). Aside from symbolizing the triumph of Western capitalism over the ideology of the former Soviet Union, the slightly comic ad also reflected the widespread assumption that the spread of market liberalization and greater political liberty went hand in hand. The ad depicts the former General Secretary of the Communist Party of the Soviet Union as he slips into a Moscow Pizza Hut for lunch with his granddaughter. There he hears the customers engaged in a lively debate about his legacy as a leader. His detractors argue that he brought economic chaos. His supporters counter that he brought opportunity. Finally, an old woman settles the argument by declaring, "Because of him, we have Pizza Hut!" This statement is roundly applauded as the patrons dig in and enjoy their meals (Stanley 1997).

The assumption that there was a natural affinity between the spread of consumer capitalism and the spread of democracy was proven fallacious, however, by trends in the former Soviet Union that were already impossible to ignore in the 1990s. Although the Russian leader Boris Yeltsin had been a passionate advocate for democracy in the face of Soviet hardliners in 1991, his prosecution of the war in Chechnya had led to him to embrace increasingly authoritarian practices by the end of that decade. Most significantly, he had paved the way for the rise of Vladimir Putin, to whom he would hand the reins of power on New Year's Eve of 1999. Putin's dual commitment to capitalism and authoritarian rule would become amply apparent during the first decade of the twenty-first century as his government silenced critics through intimidation and even, in the case of dissidents such as Anna Politkovskaya, though assassination (Kasparov 2015, p. 82).

As the world had witnessed in the response of the Chinese government to the Tiananmen Square protests of 1989, it was quite possible to embrace the liberalization of markets while brutally suppressing popular demands for liberalization in the sphere of politics. In the case of both Russia and China, of course, apologists for political repression could point to the long history of authoritarian rule in both countries, and to the threats to internal stability that motivated leaders in Beijing and Moscow to muzzle dissent. In the case of China, a long history of internal strife and foreign imperialism seemed to justify reliance on a strong centralized authority, while the government of Russia could point to the threat posed by militant Islam in regions such as Chechnya and the former Soviet republics of central Asia. After the spectacular terrorist attacks of September 11, 2001, the threat of international terrorism has prompted the erosion of civil liberties and democratic transparency throughout the industrialized world.

As Robert M. Hutchins once observed, democracy will probably not die suddenly. Instead, it faces the more insidious threat of a "slow extinction from apathy, indifference, and undernourishment" (Huntley 2001, p. 187). In this respect, the crisis of democracy in the twenty-first century has become much more serious than the mere failure of democracy to take hold in societies with a long history of authoritarian rule. Instead, the practice of democracy has come under threat in those societies, such as the United States, Britain, and France, which have seen themselves as pioneers in the establishment and promotion of democratic values. In the United States, the passage of the USA PATRIOT Act in the autumn of 2001 and the creation of the Department of Homeland Security the following year strengthened the hand of government surveillance and weakened Fourth Amendment protections for U.S. citizens in a wide variety of ways. In Britain, a pervasive system of public surveillance that had been employed to fight street crime was further expanded in the name of combating terrorism. For its part, the government of France has a long history of "asymmetric warfare" with nationalist and jihadist groups from its former colonies in North Africa, and that struggle remains a challenge to the principles of transparency and a universal respect for human rights that have long been lauded as part of the political identity of modern France. As the values of liberty, transparency, and democratic accountability have been steadily diluted in all three of these countries, a common denominator has emerged as a justification for this deliberate dilution: national security. The fact that the cause of national security has been used to justify the weakening of democracy is not surprising, but it is ironic on more than one level. On the face of it, it is plainly ironic that Britain, France, and the United States have all made democratic freedoms an essential part of their national identities, and yet the governments of all three nations have justified their attempts to roll back liberty, transparency, and accountability by appealing directly to the patriotic feelings of their citizenry. On a level that runs deeper than national identity, however, the threat posed by the trump card of national security to the fundamental values of liberalism represents a more fundamental tension within the tradition of liberal nationalism.

While most forms of nationalism have been based on appeals to a common ethnic, religious, or cultural identity, liberal nationalism derives its powers from belief in a common set of liberal ideals. One of the most concise expressions of the values of liberal nationalism in modern history is the Address that Lincoln delivered at the battlefield of Gettysburg in November of 1863. Honoring the sacrifice of those who had "given their last full measure of devotion" in order to defend the Union, the Gettysburg Address affirmed the values of a particular nation at war for its survival. And yet, in fulfilling this purpose, he transcended it. Speaking in the sparest terms possible, Lincoln identifies the Union simply as, "a new nation conceived in liberty and dedicated to the proposition that all men are created equal." His definition of the Union and its cause is in no way predicated on the religion, ethnic identity, or even the birthplace of its citizens. This creative tension between the local and the universal is part of the extraordinary power of the Gettysburg Address. While there have been many speeches and declarations celebrating such core liberal values as freedom and equality, Lincoln's address from the autumn of 1863 ties the fate of those values to the outcome of some very specific historical circumstances. The ongoing struggle to preserve the Union, Lincoln argued with his closing words, was really a universal struggle to ensure that "government of the people, by the people, and for the people shall not perish from the earth." In Lincoln's line of reasoning, the relationship between this particular nation state and the universal ideals that it represented was entirely reciprocal. The Union derived its dignity and purpose from those ideals, while the ideals themselves depended for their survival on the continuing capacity of the Union to endure.

In the most basic sense, the outcome of the American Civil War diminished the rights of states and increased the rights of individuals. As Lincoln came, somewhat tardily, to conceive of it, the most important goal of the war to eliminate the "right" of any state to practice slavery. This elimination of a "right" previously possessed by individual states was the only way to guarantee the basic rights of life and liberty to individual human beings born within the borders of the United States. This setback for *state* sovereignty was an essential advance for *individual* sovereignty, which can be the only concrete foundation for democracy. That this effort was begun imperfectly and has suffered many setbacks over the course of time should not diminish its legal and philosophical significance. Nor should it diminish its environmental significance. If we accept the premise, advanced so eloquently by Judith Shapiro, that the protection of human rights and ecological integrity are inextricably linked, then we will recognize that global democracy is essential to global sustainability. As with the abolition of slavery, the abolition of autocracy will involve an increase in the sovereign rights of individuals, and a decrease in the sovereign rights of nation states. If global democracy is ever achieved, it will not be a process of greater centralization, but of decentralization. Sovereignty will be decreased on the level of the national government, and increased on the level of the citizen.

One of the great spying scandals of the early twenty-first century illustrates the enduring tension between the rights of individuals and the strategic interests of sovereign nation states. In 2013, the former contractor Edward Snowden claimed that his primary objective in releasing a vast trove of classified data from the U.S. National Security Agency was a desire to expose the threats posed to individual liberty by the American national security state. Ironically, Snowden concluded that he could only evade prosecution in the United States by accepting the shelter offered by Vladimir Putin, who operates one of the most powerful and ruthless national security states in the world. Whatever the merits of Snowden's claims, or the wisdom of his decision to accept asylum from Putin's regime, there is at least one lesson to be drawn from his adventures in 2013. Competing national security states, far from providing a necessary check on each other's abuses, tend to exacerbate those abuses.

For all the fear that a supranational organization such as the United Nations might inspire, it is reasonable to say that on this point it has not been quite powerful enough. Although not all of the facts are known to the public at this point, let us assume that Edward Snowden was motivated by conscience to release NSA data, and was not working in the service of one state against another. If there were in fact a supranational authority capable of offering protection to an individual such as Snowden, the outcome of his case might not have strengthened the strategic position of one national security state while doing almost nothing to check the abuses of the other. If a supranational institution existed to protect whistleblowers from arrest, it would have sent a message of caution to the overweening national security apparatus of the U.S. while denying any advantage to the national security apparatus of Putin's regime. A supranational democracy could one day provide the sort of protection to whistleblowers that no single state, locked in competition with its geopolitical rivals, can provide today.

In fact, a supranational institution empowered to protect individuals from abuses of power exercised by national governments was precisely the vision that Walter Lippmann articulated in 1946 for the future of the United Nations. Advocating an international law against the further development of atomic weapons, Lippmann argued that the United Nations should offer asylum to any individual in any country who reported violations of the law by individual nation states: "Any individual scientist, industrialist, administrator, or official who wished to obey the law could, if his government were seeking to coerce him, claim the protection of the United Nations. If he escaped, they would give him asylum" (Masters and Way, eds., 1946, p. 191). This proposal is characteristic of Lippmann's vision of the potential for the United Nations in 1946. Pointing to the standards of international law that had been established by the Nuremberg Trials, Lippmann argued that the international control of nuclear technology offered another opportunity to establish the rule of law on a global scale. He saw that, by the precedent that they established, "Agreements of this type would use the liberty of the individual to regulate the absolutism of the national state" (Masters and Way, eds., 1946, p. 191). However, Lippmann warned that such agreements would only establish the rule of law if they applied to individuals, and not merely nation states. Alluding to the arguments of Alexander Hamilton in the *Federalist Papers*, Lippmann made the case that:

> All this will become possible only if we found the treaties we propose to ratify upon the basic principle that they prescribe rights and duties not only for states but for individuals. But if we do not introduce this ingredient, as Hamilton called it, then the agreements will not be laws. They will be declarations only. For observance will depend on the faithful performance by all sovereign states, and the enforcement upon the willingness and readiness of some states to wage total war in the name of collective security. (Masters and Way, eds., 1946, p. 195)

In his argument for the inherent potential of the United Nations to establish the rule of law on a global scale, Lippmann adhered to the same principle that Emery Reves had so brilliantly articulated before the Second World War was over: "We must accept the democratic conception that the state, created by the people, exists only to protect them, and maintain law and order, safeguarding their lives and liberty" (Reves 1945, p. 136). In other words, the state's only legitimate function is to protect the rights of the individual, whether it is a national state, or a world state.

As of the early twenty-first century, advocates for the creation of a global democracy have generally gravitated to one of three broad approaches. The first approach would be to bypass national governments and construct a new kind of global polity by joining local rather than national leaders together. The second involves creating a political union of democracies that would exemplify the advantages of democratic federalism and presumably attract new democracies into its ranks. The third approach involves the democratization of existing global institution such as the United Nations. As we consider the possible methods for building a global democracy, we have to concede that only time will tell which of them will offer a navigable way forward. Each broad strategy has its own drawbacks, and yet each offers some distinct advantages that the others lack.

Building a global democracy will require pragmatism, and the strategy of bypassing nation states and joining local leaders together has strong practical appeal. The most articulate advocate of this approach was the political theorist Benjamin Barber. In *If Mayors Ran the World*, Barber makes the case that city governments are at once more effective and more cosmopolitan than national governments (2013). In his characteristically poetic prose, Barber harkens back to the Athenian origins of democracy and makes the case that, "Cities woven into an informal cosmopolis can become, as the polis once was, new incubators of democracy, this time on a global scale" (Barber 2013, pp. 12–13). Although Barber died in 2017, he did live to see his idea for a "global parliament of mayors" become a reality in September of 2016 (Haldeman, 2016). Though this parliament only meets in an advisory capacity, it has the potential to set standards on transit, carbon emissions, and energy use that could have a real impact on climate change policy. On the other hand, as G. John Ikenberry observes, Barber's book does not adequately explain how a coalition of urban governments could address such problems as "stabilizing financial markets, promoting economic growth and equality, and protecting rights and freedoms" (Ikenberry 2014). Two other issues complicate Barber's claim that a regular parliament of mayors could become the foundation of a global democracy. The first issue derives from the simple fact that mayors in nondemocratic countries are not democratically elected, and so their participation in a global parliament would necessarily dilute rather than strengthen its democratic character. The second issue concerns the fact the rural citizens around the world cannot adequately be represented by a global parliament of mayors. Barber proposes some measures to address this problem, such as "including regions within representative districts" (2013, p. 345). However, he acknowledges that no such measure could change the fact that most rural citizens cannot, by definition, find full representation in the urban parliament that he has proposed. While it is true that more than fifty percent of the world's population now live in cities, it is dangerous for any system which purports to be democratic to leave an entire population behind by virtue of where they live. In much of the world, the economic and cultural divide between rural and urban dwellers is serious enough, without new forms of exclusion to exacerbate it. Even though Barber's proposed parliament of mayors would not be empowered to impose policies on rural regions, he acknowledges that its policy proposals, such as restrictions on automobile and truck transit for example, would still have some impact on people living in rural areas (2013, p. 197). However, even with these reservations noted, Barber's case for building new networks of communication and cooperation among the world's mayors may well yield dramatic results in the struggle to address climate change. As Barber correctly observes, "Intercity cooperation. .. offers an alternative strategy" for dealing with climate change because large urban centers "can directly affect carbon use within their domains through reforms in transportation, housing, parks, port facilities, and vehicles entirely under their control" (2013, p. 318).

The second path forward, which shows the strongest dedication to the basic principles of democracy, is the idea of a federal union of democracies. As noted earlier, this concept was privately discussed by Einstein in 1917, and publicly advocated by Clarence Streit in the late 1930s. As a long-term goal, it continues to be championed by the Streit Council in the twenty-first century. This path seems plausible, based on the precedent set by the success of democratic federalism in Europe after World War Two. In 1941, the antifascist activists Altiero Spinelli and Ernesto Rossi drafted a manifesto entitled "For a Free and United Europe" while they were being held as political prisoners on the Italian island of Ventotene (Mazower 2013, p. 407). This text, known as the *Ventotene Manifesto*, served as a major inspiration for the movement to build a European federation of democracies after 1945. The European Union has emerged as an enduring, if imperfect, monument to these aspirations. The historian Mark Mazower argues that the EU policies crafted in Brussels tend to serve the interests of transnational business and finance over the interests of ordinary citizens across Europe. Criticizing the steady "turn to neoliberalism in the late 1970s," and the way that the EU handled the mounting European debt crisis after 2008, Mazower laments that, "Spinelli's vision of an economy run in the service of human needs has thus been turned on its head" (2013, pp. 409–414). For all of its problems, nonetheless, the EU stands as the first functioning political and monetary union of democracies. Its existence and continued survival suggest that Clarence Streit was onto something when he argued that a transnational union of democracies was not a utopian pipe dream.

As with any grand plan for political integration, the path uniting democracies involves its own unique set of vexing challenges. The first among these is the fact a political union of democracies would likely be perceived as an aggressive power play by the authoritarian regimes that are not invited to join it, and thus may intensify geopolitical tensions as it expands. A second challenge stems from the possibility that authoritarian regimes would be able to interfere with a union of democracies by exacerbating divisions among its member states. Finally, a third potential challenge arises from the danger that such a union could fall apart through a gradual process of secession by its member states. The European Union has shown some vulnerability to all of these pitfalls. The nearest authoritarian regime, the Russian Federation, has perceived the EU as a danger to its interests, and has made negative propaganda about the EU to justify its expansionist activities, especially in Ukraine (Gessen 2017, p. 421). Confirming the second pitfall, outside forces, from both Russia and more recently from ultranationalist forces within the United States, have worked to divide the European Union and openly rooted for its demise. Finally, the success of the June 2016 British voter referendum to leave the European Union (a.k.a. the "Brexit" referendum) has now shown that the EU is indeed vulnerable to the secession of its member states.

Aware that the twenty-first century offers its own set of challenges to the idea of democratic federalism, the diplomat and scholar James R. Huntley has taken the principles first championed by Clarence Streit in 1939, and applied them to the challenges of the post-Cold War international landscape. In his book *Pax Democratica*, Huntley argues that democracy triumphed over the challenges of fascism and communism in the twentieth century as a consequence of a unique set of attributes shared by no other political system. During a period that he classes as "one of the most murderous and chaotic centuries in history" Huntley maintains that "humankind was saved by the creativity, resilience, and inherent humanity of democracy" (Huntley 2001, p. 13). Defining democracy as a political system that incorporates universal suffrage and regular multiparty elections with guaranteed civil rights and an independent judiciary, Huntley makes that case that societies that share these political values constitute a vital community that has grown considerably over the course of the past century (2001, p. 7).

Huntley's vision of democratic federalism also addresses some of the criticisms Mark Mazower has articulated concerning the neoliberal takeover of institutions such as the European Union. Although the spread of market-based economies and the globalization of trade has often been depicted as synonymous with the expansion of democracy, Huntley argues that this is not necessarily the case: "A frequent mistake is to equate democracy with capitalism. They often go together but are not the same." Offering a thumbnail definition of capitalism, Huntley highlights the fact that his book will bracket economic questions and focus exclusively on the merits of democracy as a political system: "Capitalism is another name for a free market economy in which business is largely in private hands. This book emphasizes the superiority of democracy as a way of organizing political life, including making decisions about the framework in which the economy operates, but also about many other matters" (Huntley 2001, p. 7–8). The distinction that Huntley draws here, and his decision to privilege the spread of democracy over any economic doctrine is significant. While those on the left might like to see a political revolution that would establish socialistic economic policies and those on the right might wish for a revolution that would establish laissez faire economic policies, democratic elections offer the only way that any kind of revolution can be achieved while preserving the rule of law and avoiding bloodshed. Huntley's conviction that spreading democracy is a prerequisite over all other types of fundamental change seems to affirm G. K. Chesterton's observation that "You can never have a revolution in order to establish a democracy. You must have a democracy in order to have a revolution" (Chesterton 1910, pp. 94–95).

Huntley's *Pax Democratica* outlines his vision for an institutional framework that could strengthen cooperation among democracies across the world over the course of this century. He calls for an Intercontinental Community of Democracies that would be "an expandable framework agreement" among democratic nations, that could serve as a first step to greater integration (Huntley 2001, p. 48). One of the key functions of such a cooperative framework, Huntley argues, would be environmental protection and the promotion of sustainable development (2001, p. 49). Given its pervasive emphasis on democratic federalism, it should not be surprising that Huntley's book is dedicated to Clarence K. Streit "who ushered in the era of unifying the democracies" and to EU pioneer Jean Monnet "who showed Europe how to begin" (2001, p. v).

The historian Joseph Preston Baratta, in his unparalleled two volume study of world federalism, reaches the conclusion that the facts have more than vindicated the lifelong conviction of Clarence Streit that a union of democracies must be the first step to the creation of any legitimate world federation. As Baratta puts, "A working union of peoples – as opposed to an association of states – must be based on liberal democracy first, history concludes" (2004, p. 529). Since Clarence Streit's death in 1986, the Streit Council has kept his vision of uniting the democracies alive, and signaled its relevance to contemporary challenges, including protection of the environment. The executive director of the Streit Council, Tiziana Stella stresses this connection:

> The interdependencies created by climate change . . . extend well beyond national borders. Without an overarching governance framework everyone on the planet will see their freedoms reduced – whether it's through, for example, tighter resource constraints, greater political instability, or more conflict. This governance framework could take the form of a non-binding accord like the Paris Agreement, but Trump's recent move to withdraw reveals how vulnerable this approach is to free-riding by state actors. To obtain a sustained commitment to addressing this challenge in the long-term, a binding and enforceable governance framework is needed. Its power – just as is the case with national governments – can only be made legitimate, and therefore durable, if it is based on republican principles (i.e. the principles of Union). (Stella 2017)

Tiziana Stella's observation about the current weakness of the Paris Agreement in some ways echoes Clarence Streit's observation about the weakness of the League of Nations in the decades following World War One. Just as Streit argued in the 1930s that a loose confederation of nations could not prevent aggressive wars launched by the sovereign states of Germany, Japan, and Italy, the Streit Council takes the position today that nonbinding arrangements among soverign states, such as the Paris Agreement, are not adequate to meet the challenge of climate change. Only the expansion of democratic federalism can create a "legitimate, and therefore durable" framework of "binding and enforceable governance."

Finally, the most inclusive proposal for creating democratic institutions on a global scale is the idea for a UN Parliamentary Assembly (UNPA) championed by the British journalist George Monbiot and further refined by the activist Andreas Bummel. As he works to promote the campaign for a UNPA, Bummel argues that, "If we want to strengthen the UN, the world organization at the same time needs to become more democratic" (Bummel 2017). He was first inspired to pursue this idea when the Czech playwright, political dissident, and statesman Vaclav Havel advocated the creation of a global parliament at the United Nations Millennium Summit in 2000 (Bummel 2017). In his address to the UN Millennium Summit, Havel argued for the creation of a chamber of the United Nations that would be "elected directly by the globe's population" so that the UN could become "a platform of joint, solidarity based, decision-making - by the whole of humankind - on how best to organize our stay on this planet" (Havel 2000). Inspired by this vision, Bummel has devoted himself to making Havel's words into reality, though he cautions that it will take time: "Global democracy in the sense of a citizen-elected world parliament that is able to adopt binding world law to deal with global issues is still far in the future" (Bummel 2017). In spite of the scale of this project, Bummel shares the conviction of other activists that the idea global democracy has become more plausible with advances in communications technology. Striking a similar note as the American internet activist Peter Schurman, Bummel argues that, in spite of its pitfalls, "there is no question, however, that the internet is a key factor for the development of a global consciousness."

Though ambitious, the idea of a democratic United Nations was seen as plausible by many after World War Two. In 1946, Walter Lippmann observed that "The world state is inherent in the United Nations as an oak tree is in an acorn" (Masters and Way, eds., 1946, p. 205). If the UN survives and continues to grow into the world state that Lippmann envisioned, the parallel democratization of both its institutions and member states could offer a steady path to global democracy. Because this option involves creating a global government from which no kind of state is deliberately shut out, it would be far less likely to polarize international relations than an exclusive union of democracies. On the other hand, authoritarian leaders within their given nation states could well subvert the progress of such a global institution toward the creation of global democracy. They could do this openly, by fanning the flames of nationalism within their own borders to justify their authoritarian rule. Or, they could do this covertly, by paying lip service to democratization while subverting it from within with their policies. For example, a government skilled at holding fake elections could send representatives to a world parliament that would not really represent their constituents at home, but rather the state itself. Such an outcome could easily be worse than no parliament at all. For this reason, it would probably be wise to begin reform of the United Nations by creating a "caucus of democracies" within the General Assembly, and then build the framework for a parliamentary assembly among the members of that caucus (Huntley 2001, p. 153).

As David Runciman has observed, there is a baseline level of economic prosperity that indicates whether any given society is likely to remain a democracy. Citing data accumulated over the course of the past century, Runciman concludes that, "No democracy has reverted to autocratic government once per capita GDP has risen above $7000" annually (Runciman 2013, p. 299). While it would not be safe to regard this historical precedent as establishing some sort of ironclad law, it does point to a strong correlation between economic stability and the ability of a society to remain a democracy. Given this correlation, the challenges of poverty and economic inequality throughout the world constitute perhaps the greatest single challenge to the establishment of democracy on a global scale. Particularly concerned with the economic disparities perpetuated by debt across the developing world, George Monbiot has placed a great deal of focus on changing the fundamental structures and policies of the World Bank and the International Monetary Fund. Reviewing the history of the Bretton Woods conference that laid the foundation for these institutions, Monbiot advocates the implementation of a system for international credit that was proposed by the British economist John Maynard Keynes, but scuttled by the senior American official at the conference, Harry Dexter White. While the current structure of the International Monetary Fund allows developed nations to accrue huge lending surpluses while developing nations become buried in insurmountable debt, the proposal that Keynes outlined would have used a "Clearing Union" to impose an annual limit on surplus lending and avoid the accrual of very large surpluses of very large debts by any single nation. As Monbiot sees it:

> The gift which Keynes has offered us, and which we have so far refused to accept, is a world in which the poor nations are neither condemned to do as the rich nations say, nor condemned to stay poor. A Clearing Union releases weak nations from the deficit trap, in which they must seek to produce an ever greater volume of exports in the hope of generating a sustained trade surplus, even while other, more powerful nations are trying to do the same. It ensures that demand for their exports is mobilized when it is most needed, and that nations are obliged to cooperate. Instead of seeking to beggar each other by simultaneously pursuing a trade surplus, those nations in surplus will, if the mechanism works, voluntarily go into deficit just as the deficit nations go into surplus. Instead of demanding an impossible world, the Escher's staircase envisioned by the IMF and World Bank, in which all nations simultaneously outcompete all others, it recognizes that balance can be achieved only if some nations' trade accounts descend while others rise. (Monbiot 2003, pp. 171–172.)

While the proposals that Keynes brought to Breton Woods may seem utopian in retrospect, they found vocal support among a number of maintream British economists at the time. Lionel Robbins, who headed the economics program at the London School of Economics, declared that Keynes ideas produced "an electric effect on thought" among his colleagues in the British government because "nothing so imaginative and so ambitious had ever been discussed as a possibility of responsible government policy. .. it became as it were a banner of hope, an inspiration to the daily grind of war-time duties" (Monbiot 2003. p. 164.) When these Keynesian proposals came to nothing at Breton Woods, many saw that the debt regime advocated by Harry Dexter White would lead to future disasters that would likely be avoided under the Keynes plan. The editor of *The Economist*, Geoffrey Crowther, flatly declared, "Lord Keynes was right. .. the world will bitterly regret the fact that his arguments were rejected" (Monbiot 2003, pp. 168–169.)

In addition to reforming the IMF along Keynesian lines, Monbiot also argues for the democratization of the United Nations, through the addition of a parliamentary assembly and the elimination of the veto power for permanent members of the Security Council. As is often the case with arguments for radical reform, Monbiot's critique of the present order is precise and difficult to refute, while his proposal for what should replace that order is less precise and more difficult to imagine. In his critique of the present UN Charter, Monbiot takes particular aim at the rules which govern the Security Council, especially the veto power held by its five permanent members, the United States, the United Kingdom, Russia, France, and China. It is here that democratic pretensions of the United Nations, were as Monbiot argues with eloquence, deliberately thwarted by its founders: "The Security Council mimics the notional constraints of the democratic state. By this means it claims to sustain a world order founded on right rather than might." Unfortunately, because of the veto power wielded by its five permanent members, the UN only perpetuates an international order in which "those with the might decide what is right" (Monbiot 2003. p. 69) Although Monbiot recognizes that the veto power was designed to prevent the permanent members from either leaving the UN or resorting to war with each other, he argues that the costs, especially those borne by small nations at the hands the major powers, have been too high. He reasons that, "while the veto may have functioned as a safety valve, preserving global peace at the expense of the weaker states being threatened or attacked by one of the permanent members, it has also been an instant recipe for the abuse of power and the impediment of justice" (Monbiot 2003. p. 70.)

When Monbiot wrote these words, the most recent example of the abuse of power was the illegal activities of the United States under the Bush administration. Because the United States was a permanent member of the UN Security Council, it was impossible for that body to pass any resolution condemning its invasion of Iraq, its violation of the Geneva Accords by holding prisoners indefinitely without charges at Guantanamo Bay and other U.S. military and CIA sites throughout the world, and its use of torture on prisoners at the hand of both U.S. military personnel and private contractors employed by the U.S. government. On multiple fronts, these actions marked a recrudescence of precisely the sort of militarism and lawlessness that the UN had been created to prevent, and yet the veto power possessed by the United States rendered its government impervious to any official accountability for any of these actions. Of course, the veto power possessed by other Security Council members, including vocal opponents of the Iraq war such as France, meant that the United States was unable to attain the approval the Security Council for its invasion of Iraq. In a sense, this state of affairs rendered the UN Security Council an object of abuse from all quarters. For those in the U.S. and Britain who supported the invasion of Iraq, the Security Council had failed to stand up to Saddam Hussein by not condoning the U.S. invasion of Iraq. For those who opposed that invasion, the UN Security Council had failed to prevent an unnecessary, costly, and destabilizing war by failing to stand up to the United States.

Although the various contemporary proposals for the expansion of democracy beyond national borders are quite different, it is plausible that some of them could be pursued and evaluated simultaneously. It would be possible to have a regular assembly of mayors, as well as a regular assembly of elected representatives from around the world. Even a union of democracies, though it would be distinct from these other bodies, could operate alongside them, much as current IGOs, such as OECD or the Commonwealth of Nations operate alongside (and overlap with) such organizations as the UN or G20. In a similar manner, the global parliament of mayors inspired by Benjamin Barber could be combined with the principle of democratic federalism championed by Clarence Streit. Urban governments confront issues of energy and environmental stability every day, and the principle of democratic federalism allows individual political units to be laboratories for political, economic, and social experiments. Restrictions on motor vehicle traffic, carbon taxes, and mixed used neighborhoods that cut down on long commutes are all examples of urban experimentation and each of them is relevant to climate change. One way to integrate the principle of democratic federalism with a parliament of mayors would be to create a caucus of democratically elected mayors within that parliament. As part of a caucus of democratic cities, municipal governments could exchange ideas and provide support for each other on an international stage. In the case of cities that face extreme pressure from the national governments, such as Hong Kong, membership in a caucus of democratic cities could provide useful support for local autonomy and the preservation of democratic institutions. In whatever order they are pursued, the principle that should dominate the implementation of any of these proposals is the strengthening of cooperative ties between democratic governments with the aim of creating global institutions that are in themselves democratic. This was a principle that Clarence Streit understood better than most. If we seek to create a global democracy, it is wiser to build democratic institutions first, and expand them until they are global, than it is to create global institutions first in the hope that they will one day become democratic. For this reason, Streit's principle of unifying existing democracies is probably the surest path to the creation of global democracy.

As noted in the introduction, the establishment of democracy beyond the boundaries of the nation state is not in itself the solution to the challenges posed by climate change. It is, however, a fundamental *prerequisite* to any viable and lasting solution. National governments and multinational corporations must live under the rule of law, not according to agreements that any party can abandon at will. And the rule of law, in order to be accepted as legitimate on the time scale necessary to cope with climate change, must be democratic in its origins. That said, the most vexing problem is still how we get there from here. Global democracy remains a dream, but the history of a dream can teach us something. Da Vinci's sketches of flying contraptions do not closely resemble the real flying machines that emerged more than half a millennium later, but they do exemplify the same combination of bold imagination and sober discipline that would change the world at Kitty Hawk. In a similar way, the rough sketches of global democracy that dreamers such as Alfred Tennyson, H. G. Wells, and Albert Einstein left behind reveal, for all their flaws, the tantalizing outlines of what is possible. Although Einstein was routinely dismissed as naïve when he advocated global democracy, his meditations on how it might be possible are infused with the same sort of genius that make Da Vinci's sketches of flying machines so beautiful and prophetic.

In the 1840s, when Tennyson dared to imagine a "parliament of man" and a "federation of the world" in his poem "Locksley Hall" he outlined a vision that, like the British Empire itself, was at once expansive and parochial. In the same poem, Tennyson's dismissal of East Asian history and culture is expressed in the line, "Better fifty years of Europe than a cycle of Cathay." While the pithy chauvinism of this line may have been immensely satisfying to British readers in the age of Queen

Victoria, it reflects the stark limits of the poet's vision. A viable federation is only possible among equals, so the inequality inherent in imperialism cast a shadow over real attempts to create a world federation in the twentieth century, such as the League of Nations and the UN. Because colonial powers held the top positions in both of these organizations, they had far less credibility among people subjected to colonial rule. While two world wars exposed the stark limitations of nationalism among industrialized nations, the long-term effects of racism and foreign domination made nationalism a powerful idea across Asia, Africa, and every region where people sought to free themselves from colonialism.

On the eve of World War One, H. G. Wells sketched his own vision of an enlightened and democratic world state in his novel *The World Set Free*. As time progressed, however, Wells increasingly distrusted the ability of the *demos* to understand the technological complexities of the modern world and avoid further catastrophes engendered by atavistic nationalism, craven superstition, and greed. His later visions of a desirable future, sketched in such novels as *The Shape of Things to Come*, were committed to a technocratic, rather than democratic, vision of the world state. Like the Guardians of Plato's *Republic*, a superior and selfless class of scientists and engineers was required to create and maintain the ideal state the Wells envisioned. Given the man-made disasters that tore the world apart in his lifetime, it is perhaps not surprising that Wells put his faith in technocracy rather than democracy (Deudney 2007, pp. 207–208). He was not alone. In the early years of the Great Depression, in fact, Technocracy became a full-blown political creed, especially in the United States. There are two serious problems with this way of approaching global governance. First, such thinking confuses ends and means. Inventors, engineers, and managers are excellent at discovering the *means* to do something, whether it is building an atomic bomb, constructing a continental network of ferroconcrete multilane highways, or sending people to the moon. Decisions about what *ends* to pursue, however, remain firmly embedded in the realm of politics, and one still must decide whether those politics are going to be conducted along authoritarian or democratic lines. The second problem with technocracy was pointed out by Bertrand Russell when he critiqued the modern temptation to view governing as an engineering project, and to view the people themselves as the "raw material" for that project (Russell 1922, p. 81).

Like Wells, Albert Einstein loathed militaristic nationalism and began exploring the possibility of democracy on a global scale during the First World War. The vision that Einstein sketched in a letter to a friend in August of 1917 was more detailed than the visions of either Tennyson or Wells. However, Einstein's faith in the possibility of a democratic future for the human race oscillated with the peaks and troughs of democratic fervor in the twentieth century. In the summer of 1917, newspapers across the world hailed Kerensky as the man who could make Russia into a constitutional democracy (Runciman 2013, pp. 40–41). Excited by the prospect of democracy spreading among the Allied powers that summer, Einstein dreamed of a union of democracies with the United States, Britain, France, and Russia at its helm (Rowe and Schulman 2007, pp. 77–78) Einstein's vision of global democracy was refined in the decades before World War Two through a long correspondence with the journalist and democratic activist Emery Reves. The two men had met in Geneva in 1932 and shared a powerful commitment to democracy and antifascism. Reves shared Einstein's conviction that democracy needed to transcend the nation state, and articulated that vision in two books published during the Second World War, *A Democratic Manifesto* (1942) and *The Anatomy of Peace* (1945). Inspired by the principles that Reves had outlined in these books, Einstein invested his reputation in the cause of cosmopolitan democracy after World War Two. In addition to endorsing *The Anatomy of Peace*, Einstein began to air his own ideas about what a democratic world state might look like. In his conversations with the broadcast journalist Raymond Swing, Einstein even proposed global voting franchise that would bypass individual national governments and elect a global parliament that would only be accountable to voters, and not national governments.

By the year of his death, however, Einstein had trimmed his sails considerably. The explosion of hydrogen bombs by both the United States and the Soviet Union in the early 1950s had convinced many observers that the very survival of the human race was under threat from the current arms race. In 1955, Einstein joined forces with Bertrand Russell to issue a warning about the dangers posed by the hydrogen bomb, and a non-ideological for plea for peace and cooperation across national boundaries. From the vantage point of 1955, the ideals of democracy had lost none of their luster, but they would mean nothing if the human race were to commit collective suicide with thermonuclear weapons. Einstein had not abandoned the cause of expanding democracy beyond the nation state, but he left that struggle to future generations as he tightened his focus on those goals that he saw as most urgent at the time, especially arms control and international cooperation among scientists. The ambitious vision of the young Einstein, who dreamed of a global union of democracies, had been put on the shelf so that he could concentrate on the cause of helping humanity to survive another day.

In the early twenty-first century, we may find that there is something useful to be learned from the more ambitious and democratic musings of the younger Einstein. Having survived the Cold War arms race, but facing a new threat in the form of climate change, we can see that near term and long-term goals are mercurial things and prone to trading places. Where we once needed to assure our survival in order to expand democracy, we must now expand democracy in order to assure our survival. Unless we bring the responsiveness, adaptability, and accountability of democracy to the global regulation of trade and industry, we will not be able to deal with climate change and the myriad disasters and disruptions that it is bound to engender. If each democracy attempts to weather the coming storm in the service of its own "national interest," none of them may survive, at least not in the form that we might honestly describe as a democracy. Such democratic principles as due process, privacy, freedom of the press, and free and fair elections have already been eroded in the twenty-first century, and they are not likely to survive in a world where extreme weather events, droughts, famines, and mass migrations are addressed by an anarchic society of sovereign nation states, each angling for its own advantage in a zero-sum game. Conversely, if we can extend these vital principles of democracy beyond the nation-state, we will increase our own chances for survival through rational, accountable, and flexible cooperation on a global scale.

The most difficult question regarding global democracy is not whether we should have it, but how we could possibly achieve it. The initiative in this case will not come from governments but from private citizens joining forces across national borders. The abolition of slavery in the nineteenth century and the political enfranchisement of women in the twentieth century both furnish excellent examples of how movements by individual citizens can lead to fundamental social, economic, and political change on a global scale. In his essay "The State of the Species" the journalist and author Charles Mann describes the scale of tremendous behavioral changes that have taken place in the past two centuries, including the statistical decline in violence documented by Steven Pinker, the near total eradication of slavery, and the growing enfranchisement of women across the world. Mann attributes these dramatic changes to the "behavioral plasticity" of human beings, "a defining feature of *Homo sapiens'* big brain." Citing more quotidian examples, Mann observes that this plasticity "means that humans can change their habits; almost as a matter of course, people change careers, quit smoking or take up vegetarianism, convert to new religions, and migrate to distant lands where they must learn strange languages" (2012). While it is far from inevitable that we will change our collective behavior soon enough to avoid catastrophic climate change, Mann submits that it is at least a possibility.

Pointing to the vast human potential that has been liberated by social progress of the that past two centuries, Mann observes that, "removing the shackles from women and slaves has begun to unleash the suppressed talents of two-thirds of the human race. Drastically reducing violence has prevented the waste of countless lives and staggering amounts of resources." He then poses the rhetorical question of whether "we wouldn't use those talents and those resources to draw back before the abyss?" Of course, the jury is still out on whether past successes in human progress that Mann discusses portend future success in addressing the unprecedented challenge of climate change. However, Mann's point about liberating "the suppressed talents of two thirds of the human race" suggests that supranational democracy is the political system most suited to meet that challenge. The facts on the ground indicate that the protection of individual rights and access to education for women can pay dramatic dividends in fighting climate change. Paul Hawken reports that educating girls and women is "the most powerful lever available for breaking the cycle of intergenerational poverty, while mitigating emissions by curbing population growth" (2017, p. 81). Hawken also cites research that ranks campaigns to educate girls, such as those led by Malala Yousafzai of Pakistan, as among the most cost-competitive options for reducing carbon dioxide emissions, requiring an approximate investment of "just ten dollars per ton of carbon dioxide" (2017, p. 81).

In the years following the Civil War, Walt Whitman penned an essay called "Democratic Vistas" in which he identified the creation of a universal community that honored each individual as the ultimate goal of democracy. This was a powerful ethos that "ever seeks to bind, all nations, all men, of however various and distant lands, into a brotherhood, a family." Whitman identified this audacious goal as "the old, yet ever-modern dream of earth, out of her eldest and her youngest, her fond philosophers and poets." Though Whitman viewed the horizontal expansion of democracy as encompassing the whole human race, he viewed the powers of any democratic government as limited by a necessary respect for the autonomy and responsibility of the individual. As Whitman saw it, the "mission of government" was "to train communities through all their grades, beginning with individuals and ending there again, to rule themselves" (Whitman [1871], 1982. p. 935).

The relationship between a butterfly and its chrysalis offers a biological analogy that could shed some light on this relationship between the ideals of democracy and the sheltering institutions of the nation state. In the closing words of his Gettysburg Address, Lincoln alluded to the broader significance of the struggle to preserve the Union for the fate of democracy across the world. In Lincoln's reasoning, the function of the Union was not only to protect the rights of its citizens but also to provide a shelter for a democratic movement that transcended national borders. Like a chrysalis defending the slow transformation of a caterpillar into a butterfly, the Union provided an irreplaceable shelter in which the culture and legal institutions of the United States could mature into a new kind of democracy: "a new birth of freedom" that would be unprecedented in size and scope. The analogy of the chrysalis and the butterfly, like so many analogies drawn from nature, entails both creation and destruction. For the butterfly to take flight, it must tear open the shelter of the chrysalis and leave it behind. What had been a shelter would become a sarcophagus if this process did not take place. The nation state, which has sheltered democracy for centuries, will become its sarcophagus if democratic institutions are not allowed to grow and evolve beyond the narrow scope of national borders.

The question that the human race faces in the twenty-first century is not whether we should or should not have global governance. The global governance that we already have insures the nearly frictionless flow of goods and services around the planet by maintaining and expanding a transport and communications infrastructure that dwarfs anything seen in all of human history. The real question is whether we can make the global governance that we already have fairer, more democratic, and more effective in protecting the lives and wellbeing of the living and the yet to be born.

## Bibliography

Barber, Benjamin. 2013. *If mayors ruled the world: Dysfunctional nations, rising cities*. New Haven: Yale University Press.

Blom, Philipp. 2008. *The vertigo years: Europe, 1900–1914*. New York: Basic Books.

Bumell, Andreas. 2017. Personal interview via email. 29 May 2017.

Chesterton, G.K. 1910. *Tremendous trifles*. New York: Dodd, Mead and Company.

Deudney, Daniel H. 2007. *Bounding power: Republican security theory from the polis to the global village*. Princeton: Princeton University Press.

Ferris, T. 2011. *The science of liberty: Democracy, reason, and the laws of nature*. New York: Harper Perennial.

Gessen, Masha. 2017. *The future is history: How totalitarianism reclaimed Russia*. New York: Riverhead Books.

Gilbert, Marin. 1997. *Winston Churchill and Emery Reves: Correspondence, 1937–1964*, 1997. Austin: University of Texas Press.

Havel, Václav. 2000. "Address by Václav Havel, President of the Czech Republic at the Millennium Summit of the United Nations" New York, September 8th, 2000. http://www.vaclavhavel.cz/showtrans.php?cat=projevy&val=70_aj_projevy.html&typ=HTML (Accessed 29 Aug 2017).

Hawken, Paul. 2017. *Drawdown: The most comprehensive plan ever proposed to reverse global warming*. New York: Penguin Books.

Hesiod. 2017. *The poems of Hesiod: Theogony, works and days, and the shield of herakles*. Trans. Barry B. Powell. Oakland: University of California Press.

Holden, Barry, ed. 2000. *Global democracy: Key debates*. London: Routledge.

Huntley, James R. 2001. *Pax democratica: A strategy for the 21st century*. Palgrave.

Ikenberry, G. John. 2014. Review of *If Mayors Ruled the World* by Benjamin Barber. *Foreign Affairs* 93(1). January–February, 2014.

Ikenberry, H. John, and Anne-Marie Slaughter. 2006. *Forging a world order of liberty under law: Final paper of the Princeton project on national security*. Princeton: The Woodrow Wilson School of Public and International Affairs at Princeton University.

Kasparov, Garry. 2015. *Winter is coming: Why Vladimir Putin and the enemies of the free world must be stopped*. New York: Public Affairs Books.

———. 2017. "Donald's Pravda: Trump and his apologists spookily echo Vladimir Putin" New York Daily News. Sunday, July 16, 2017.

Libby, Leona Marshall. 1971. *Technological risk versus natural catastrophe (P-4602)*. Santa Monica: RAND Corporation.

Lippmann, Walter. 2007a. *"International control of atomic energy" in one world or none* [first published in 1946], 205. New York: The New Press.

Lippmann, Water. 2007b. "International Control of Atomic Energy" in in *One World or None*. [first published in 1946] (New York: The New Press, 2007), p. 191, 195, 205.

Lynch, M. 2001. Pandora's ballot box: Comments on the 2000 US presidential election. *Social Studies of Science* 31 (3): 417–419. Retrieved from http://www.jstor.org.ezproxy.bu.edu/stable/3183007.

Mann, Charles C. 2012. "The State of the Species" *Orion Magazine*. October 24th, 2012. https://orionmagazine.org/article/state-of-the-species/ (Accessed 29th Aug 2017).

Mazower, Mark. 2013. *No enchanted palace: The end of empire and the ideological origins of the United Nations*. Princeton: Princeton University Press.

Monbiot, George. 2003. *The age of consent*. London: Flamingo. 164, 168–169, 171–172, 69.

Monnet, Jean, and Richard Mayne. trans1978. *Memoirs*. New York: Doubleday & Company.

Pinker, Steven. 2018. *Enlightenment now: The case for reason, science, humanism, and progress*. New York: Viking.

Reves, Emery. 1945. *The anatomy of peace*. New York: Harper and Brothers.

Ricks, Thomas. 2017. *Churchill and orwell: The fight for freedom*. New York: Penguin Press.

Rowe, David, and Robert Schulmann, eds. 2007. *Einstein on politics*. Princeton: Princeton University Press.

Runciman, David. 2013. *The confidence trap: A history of democracy in crisis from world war I to the present*. Princeton: Princeton University Press.

Russell, Bertrand. 1922. *The problem of China*. New York: Century Co.

Stanley, Alessandra. 1997. "From Perestroika to Pizza." The New York Times. December 3rd, 1997. http://www.nytimes.com/1997/12/03/world/from-perestroika-to-pizza-gorbachev-stars-in-tv-ad.html

Streit, Clarence K. 1939. *Union now: A proposal for a federal union of the democracies of the North Atlantic*. London: Jonathan Cape.

Suskind, R. 2004. Faith, certainty and the presidency of George W. Bush. New York Times, p. 17.

Tägil, S. 1998. *Alfred Nobel's thoughts about war and peace*. Norway: Nobel prize org.

Whitman, Walt. 1982. *Walt Whitman: Poetry and prose*. New York: Library of America.

# Master Bibliography


Abrams, I. 2005. *Bertha von Suttner and the Nobel Peace Prize*. Hague: Peace Palace Library.

Akin, William. 1977. *Technocracy and the American dream: The American technocrat movement, 1900–1941*. Berkeley: University of California Press.

Almond, Philip C. 1999. *Adam and eve in seventeenth century thought*. Cambridge: Cambridge University Press.

Anderson, Benedict. 1983. *Imagined communities: Reflections on the origin and spread of nationalism*. New York: Verso.

Anderson, Larry. 2002. *Benton MacKaye: Conservationist, planner, and creator of the Appalachian trail*. Baltimore: Johns Hopkins University Press.

Angell, Norman. 1913. *The great illusion*. New York/London: G. P. Putnam's Sons.

Appiah, Kwame Anthony. 2006. *Cosmopolitanism: Ethics in a world of strangers*. New York: W.W. Norton & Co.

Archibugi, Daniele, and David Held. 1995. *Cosmopolitan democracy: An agenda for a new world order*. Oxford, UK/Cambridge, MA: Polity Books.

Ash, Timothy Garton. 2008. We friends of liberal international order face a new global disorder. *The Guardian*. September 10th, 2008.

Atknison, David C. 2016. *The burden of white supremacy: containing Asian migration in the British Empire and the United States*. Chapel Hill: The University of North Carolina Press.

Bacevich, Andrew. 2008. *The limits of power: The end of American exceptionalism*. New York: Henry Holt & Company.

Bacon, Francis. [1620] 2000. *The new Organon: Cambridge texts in the history of philosophy*. Cambridge: Cambridge University Press.

Bacon, F., L. Jardine, and M. Silverthorne. 2000. *Francis Bacon: The new Organon*. Cambridge, MA: Cambridge University Press.

Bailey, Ronald. 1994. *Ecoscam: The false prophets of ecological Apocalypse*. New York: St. Martin's Press.

Baratta, Joseph Preston. 2004a. *The politics of world federation*. Vol. 1. Westport/Connecticut: Praeger.

———. 2004b. *The politics of world federation*. Vol. 2. Westport/Connecticut: Praeger.

Barber, Benjamin. 2013. *If mayors ruled the world: Dysfunctional nations, rising cities*. New Haven: Yale University Press.

Bauer, Ela. 2016. Jan Gottlieb Bloch: Polish cosmopolitanism vs. Jewish universalism. In *Cosmopolitanism, nationalism, and the Jews of East Central Europe*, ed. Michael L. Miller and Scott Urey. London: Routledge.

Bender, Marilyn. 1969. The Lone woman in think tank Is a 'Renaissance Man.' *New York Times*. May 27, 1969. p. 50.




R. S. Deese, *Climate Change and the Future of Democracy*, Environmental Challenges and Solutions 5, https://doi.org/10.1007/978-3-319-98307-3

Bentele, K., and E. O'Brien. 2013. Jim Crow 2.0? Why states consider and adopt restrictive voter access policies. *Perspectives on Politics* 11 (4): 1088–1116. https://doi.org/10.1017/S1537592713002843.

Berger, Ranier. 1983. "Forward" in Libby, Leona Marshall. In *Past climates: Tree thermometers, commodities, and people*. Austin: University of Texas Press.

Berlin, Isaiah, and Henry Hardy, eds. 2013. *The crooked timber of humanity: Chapters in the history of ideas*. 2nd ed. Princeton: Princeton University Press.

Bernstein, Irving. 1993. *Promises kept: John F. Kennedy's new frontier*. Oxford: Oxford University Press.

Bess, Michael. 2006. *Choices under fire: Moral dimensions of world war two*. New York: Vintage Books.

Bhagavan, Manu. 2012. *The peacemakers: India and the quest for one world*. New Delhi: Harper Collins & The India Today Group.

Blom, Philipp. 2008. *The vertigo years: Europe, 1900–1914*. New York: Basic Books.

Borgese, Elisabeth Mann. 1965. *A constitution for the world*. Santa Barbara: Center for the Study of Democratic Institutions.

———. 1969. *Lecture on the ocean regime*.

Borgese, Giuseppe Antonio, et al. 1947. Brief history of the committee. *Common Cause*, 11.

Bowman, Durrell. 2016. *Experiencing peter gabriel: A listener's companion*. New York: Rowman & Littlefield.

Boyer, John W. 1995. Drafting salvation. *University of Chicago Magazine*. Volume 88, Number 2.

Bradley, David. 1948. *No place to hide*. Boston: Little, Brown.

Brand, Stewart. 2009. *Whole earth discipline: An ecopragmatist manifesto*. New York: Viking Penguin.

Bray, Daniel, and Steven Slaughter. 2015. *Global democratic theory: A critical introduction*. Cambridge, UK: Polity Press.

Buckley, John. 2006. *Air power in the age of total war*. New York: Routledge.

Buckley, Terry. 2010. *Aspects of Greek history 750–323 BC: A Source-Based Approach*.

Bull, Hedley. 1977. *The anarchical society: A study of order in world politics*. New York: Columbia University Press.

Bumell, Andreas. 2017. Personal interview via email. May 29th, 2017.

Burke, Edmund. [1790] 2012. *Reflections on the revolution in France*. Mineola: Dover Publications, Inc.

Cadier, D., and M. Light, eds. 2015. *Russia's foreign policy: Ideas, domestic politics and external relations*. New York: Palgrave Macmillan.

Calhoun, Craig. 1997. *Neither gods nor emperors: Students and the struggle for democracy in china*. Oakland: University of California Press.

Cameron, James and Jon Landau (producers) and Cameron, James (Director). 2009. *Avatar*. United States: Twentieth Century Fox.

Carlarne, Cinnamon P. 2010. *Climate change law and policy: EU and US approaches*. Oxford: Oxford University Press.

Carrington, Damian. 2016. The Anthropocene epoch: Scientists declare the dawn of human-influenced age. *The Guardian*. https://www.theguardian.com/environment/2016/aug/29/declare-anthropocene-epoch-experts-urge-geological-congress-human-impact-earth.

Carson, Rachel. [1962] 2002. Silent spring. New York: Houghton Mifflin.

Ceadal, Martin. 2009. *Living the great illusion: Sir Norman Angell, 1872–1967*. Oxford: Oxford University Press.

Center for Health, Environment & Justice website. 2016. Accessed 23 June 2014. http://chej.org/about/mission/.

Chan, Wing-tsit. 1963. *A source book in chinese philosophy*. Princeton: Princeton University Press.

Chesterton, G.K. 1910. *Tremendous trifles*. Mead and Company: Dodd.

Clarke, Arthur C. 2001. *Greetings, carbon-based bipeds!: collected essays, 1934–1998*. New York: Macmillan.

Clayton, Bruce. 1998. *Forgotten prophet: The life of Randolph Bourne*. Columbia: University of Missouri Press.
Cleave, Mary L. 2011. Interview via email.
Cloud, Preston. 1989. *Oasis in space: Earth history from the beginning*. New York: W. W. Norton & Co.
Collier, Aine. 2007. *The humble little condom: A history*. Amherst: Prometheus Books.
Commoner, Barry. 1966. *Science and survival*. New York: Viking Press.
Congressional Budget Office (CBO). 2008. Issues and options in infrastructure investment Appendix A: Spending for research and development and for education.
Cronon, William. 1996. *Uncommon ground: Rethinking the human place in nature*. New York: W. W. Norton & Co.
Davies, S. 2016a. *Adaptable livelihoods: Coping with food insecurity in the Malian Sahel*. Springer.
Davies, Jeremy. 2016b. *The birth of the anthropocene*. Oakland: University of California Press.
de Condorcet, M. 1795. *Sketch for a historical picture of the progress of the human mind*. Trans. June Barraclough. New York: Noonday Press, 1955.
De Vitoria, Francisco. 1991. In *Political writings*, ed. Anthony Pagden and Jeremy Lawrence. Cambridge: Cambridge University Press.
Deese, R.S. 2008. A metaphor at midlife: 'The Tragedy of the Commons' turns 40. *Endeavour* 32 (4): 152–155.
———. 2009. The artifact of nature: Spaceship earth and the dawn of global environmentalism. *Endeavour* 33 (2): 70–75.
———. 2015. *We are amphibians: Julian and Aldous Huxley on the future of our species*. Oakland: University of California Press.
Deneen, Patrick J. 2014. *Democratic faith*. Princeton: Princeton University Press.
Desilver, Drew. 2017, December . *Despite concerns about global democracy, nearly six-in-ten countries are now democratic*. FactTank Blog, .Pew Research Center. http://www.pewresearch.org/fact-tank/2017/12/06/despite-concerns-about-global-democracy-nearly-six-in-ten-countries-are-now-democratic/. Accessed 5 Jan 2018.
Deudney, Daniel H. 2007. *Bounding power: Republican security theory from the polis to the global village*. Princeton: Princeton University Press.
Diamond, Jared. 1999. *Guns germs and steel: The fates of human societies*. New York: W.W. Norton & Company.
Du Bois, W. E. B. 2012. Quoted in Tuck, Steven. *Fog of war: The second world war and the civil rights movement*. Oxford: Oxford University Press, p. 202.
Duchene, Francois. 1994. *Jean Monnet: The first statesman of interdependence*. New York: W.W. Norton & Co.
Dyson, Freeman. 2008. *The scientist as rebel*. New York: NYREV, Inc.
Eaton, Casindania P., et al. n.d. Peace by Law Our One Hope. Letter to the *New York Times,* Oct 10, 1945. p. 20.
Egan, Michael. 2007. *Barry commoner and the science of survival: The remaking of American Environmentalism*. Cambridge, MA: MIT Press.
Einstein, Albert. 1950. *Out of my later years*. New York: The Philosophical Library, Inc.
Einstein, A., & Swing, R. 1947. *Atomic war or peace*. Emergency committee of atomic scientists. In Einstein, *out of my later years*. New York: The Philosophical Library, Inc. 1950.
Einstein, Albert, David Rowe, and Robert Schulmann, eds. 2007. *Einstein on Politics*. Princeton: Princeton University Press.
Eisenhower, Dwight. 1953. The chance for peace. April 16th 1952. Eisenhower Presidential Library. https://www.eisenhower.archives.gov/all_about_ike/speeches/chance_for_peace.pdf
Emerson, Sarah. 2017. "My world's on fire: We asked smash mouth if 'All Star' is about climate change" *Motherboard*. https://motherboard.vice.com/en_us/article/qkqdm7/is-smash-mouth-all-star-about-climate-change-global-warming. Accessed 24 Jan 2018.
Evans, James. 1998. *The history and practice of ancient astronomy*. Oxford: Oxford University Press.

Falk, Richard and Andrew Strauss. 2002. People want a say: Next, a global parliament. *New York Times*. April 19, 2002.

Fang, Lizhi. 1989. Keeping the faith. *New York Review of Books*. December 21st, 1989.

Ferkiss, Victor C. 1993. *Nature, technology, and society : Cultural roots of the current environmental crisis*. New York: New York University Press.

Ferris, T. 2011. *The science of liberty: Democracy, reason, and the laws of nature*. New York: Harper Perennial.

Fleming, James. 2010. Fixing the sky: The checkered history of weather and climate control. New York: Columbia University Press, p. 129, 132, 217.

Foa, R. S. & Mounk, Y. (2017). The signs of deconsolidation. Journal of Democracy 28(1), 5–15. Johns Hopkins University Press.

Foley, Michael. 2014. *Rise of the tank: Armoured vehicles and their use in the first world war*. South Yorkshire: Pen and Sword Books, Ltd.

Forrester, Jay W. 1971. *World dynamics*. Cambridge: Wright-Allen Press.

Forster, William Lloyd. 1883. *Two lectures on checks to population*, 31. Oxford/London: J.H. Parker/J.G. and F. Rivington.

Fuller, Buckminster. 1982. *Critical path*. New York: St. Martin's Griffin.

Garan, Ron. 2015. *The orbital perspective*. Berrett-Koehler Publishers.

Garten, Jeffrey E. 2016. *From silk to silicon: The story of globalization through ten extraordinary lives*. New York: Harper Collins.

Geddes, Patrick. 1915. *Cities in evolution*. London: Williams and Norgate.

Gessen, Masha. 2017. *The future is history: How totalitarianism reclaimed russia*. New York: Riverhead Books.

Gilbert, Marin. 1997. *Winston Churchill and Emery Reves: correspondence, 1937–1964*. Austin: University of Texas Press.

Gore, Al. 2007. Nobel prize lecture. https://www.nobelprize.org/nobel_prizes/peace/laureates/2007/gore-lecture_en.html. Accessed 3 Jan 2018.

Gorelik, Gennady, and Antonina W. Bouis. 2005. *The world of andrei sakharov: A russian physicist's path to freedom*. Oxford: Oxford University Press.

Grotius, Hugo. 2009. *Hugo Grotius Mare Liberum 1609–2009: Original Latin Text and English Translation*. Brill.

Haldane, J.B.S. 1924. *Daedalus, or science and the future*. London: K. Paul, Trench, Trubner & Co.

Haldevang, Max De. 2016 "Mayors for 70 global cities are gathering to try to solve the world's problems" *Quartz*, September 9th, 2016. https://qz.com/776882/mayors-from-70-cities-around-the-world-are-gathering-to-try-to-solve-the-worlds-problems/. Accessed 30 Aug 2017.

Hall, Timothy L. 2001. *Supreme court justices: A biographical dictionary*. New York: Facts on File.

Hamblin, Jacob Darwin. 2013. *Arming mother nature: The birth of catastrophic environmentalism*. Oxford: Oxford University Press.

Hamilton, J. R. 1999. *Plutarch: Alexander*. Duckworth Publishing.

Hamilton, Clive. 2013. *Earthmasters: The dawn of the age of climate engineering*. New Haven: Yale University Press.

Hansen, James. 2009. *Storms of my grandchildren: The truth about the coming climate catastrophe and our last chance to save humanity*. New York: Bloomsbury.

Hansen, James. et al. 1981. Climate impact of increasing atmospheric carbon dioxide. *Science*, August 28.

Hardin, Garrett. 1968. The tragedy of the commons. In *Managing the commons, eds.* Hardin, Garrett and John Baden, 20–24. San Francisco: W.H. Freeman and Company, (1977).

———. 1973. *Exploring new ethics for survival: The voyage of the spaceship beagle*. Baltimore: Penguin Books.

———. 1993. *Living within limits: Ecology, economics, and population taboos*. Vol. 21, 24–25. Oxford: Oxford University Press.

Hardin, G. 1999. The Persistence of the species. *Politics and the life sciences* 18 (2): 225–227. Retrieved from http://www.jstor.org.ezproxy.bu.edu/stable/4236506.

Havel, Václav. 2000. Address by Václav Havel, President of the Czech Republic at the Millennium Summit of the United Nations. New York, September 8th, 2000. http://www.vaclavhavel.cz/showtrans.php?cat=projevy&val=70_aj_projevy.html&typ=HTML. Accessed 29 Aug 2017.

Hawken, Paul. 2017. *Drawdown: the most comprehensive plan ever proposed to reverse global warming*. New York: Penguin Books.

Hesiod. 2017. *The poems of Hesiod: Theogony, works and days, and the shield of herakles*. Trans. Barry B. Powell. Oakland: University of California Press.

Heuser, Beatrice. 2010. *The evolution of strategy: Thinking war from antiquity to the present*. Cambridge: Cambridge University Press.

Hill, James. 2011. *Descartes and the doubting mind*. London: Bloomsbury.

Holden, Barry, ed. 2000. *Global democracy: Key debates*. London: Routledge.

———. 2002. *Democracy and global warming*, 6–7. London: Continuum.

Howe, Nicolas. 2016. *Landscapes of the secular: Law, religion, and american sacred space*. Chicago: University of Chicago Press.

Huntley, James R. 2001. *Pax Democratica: A strategy for the 21st Century*. Basingstoke: Palgrave.

Hutchins, Robert Maynard. 1954. *Great books: The foundation of a liberal education*, 45. New York: Simon & Schuster.

Huxley, Aldous. [1962] 2002. Island. New York: Harper Perennial Classics.

Ibsen, Henrik. [1882] 2007. An enemy of the people. Trans. Nicholas Rudall. Chicago: Ivan R. Dee.

Ikenberry, G. John. 2014. Review of *If Mayors Ruled the World* by Benjamin Barber]. *Foreign Affairs,* Volume 93, Number 1. January–February, 2014.

Inhofe, James. 2012. *The greatest hoax: How the global warming conspiracy threatens your future*. New York: WND Books.

Isaacson, Walter. 2007. *Einstein: His life and universe*. New York: Simon and Schuster.

Jakhu R.S., Pelton J.N., Nyampong Y.O.M. 2017. *Space mining and its regulation*. Springer Praxis Books. Cham: Springer.

Josephson, Paul R. 2005. *Resources under regimes: Technology, environment, and the state*. Cambridge, MA: Harvard University Press.

Jung, Carl, ed. 1964. *Man and his symbols*. Garden City: Doubleday & Company.

Kant, Immanuel. 1991. In *Kant: Political writings*, ed. Hans Siegbert Reiss. Cambridge: Cambridge University Press.

Kant, I., P. Kleingeld, J. Waldron, M.W. Doyle, and A.W. Wood. 2006. *Toward perpetual peace and other writings on politics, peace, and history*. New Haven: Yale University Press.

Kaplan, Temma. 2015. *Democracy: A world history*. Oxford: Oxford University Press.

Kasparov, Garry. 2015. *Winter is coming: Why vladimir putin and the enemies of the free world must be stopped*. New York: Public Affairs Books.

———. 2017. Donald's Pravda: Trump and his apologists spookily echo Vladimir Putin. *New York Daily News*. Sunday, July 16, 2017.

Kennan, George F. 1993. *Around the cragged hill: A personal and political philosophy*. New York: W. W. Norton.

Kennedy, John F. 1960. Democratic National Convention, 15 July 1960. John F. Kennedy Presidential Library. https://www.jfklibrary.org/Asset-Viewer/AS08q5oYz0SFUZg9uOi4iw.aspx

———. 1962. Address at Independence Hall, Philadelphia, Pennsylvania, July 4, 1962. John F. Kennedy Presidential Library. https://www.jfklibrary.org/Research/Research-Aids/JFK-Speeches/Philadelphia-PA_19620704.aspx

Kennedy, Paul. 2006. *The parliament of man: The past, present, and future of the united nations*. New York: Vintage.

Keohane, Robert O., and Elinor Ostrom. 1995. *Local commons and global interdependence: Heterogeneity and cooperation in two domains*. London: Sage Publications.

Keylor, William R. 2001. *The twentieth century world: An international history*. Oxford: Oxford University Press.

Klein, Naomi. 2015. *This changes everything: Capitalism vs. the climate*. New York: Simon & Schuster.

Koehn, P.H. 2015. *China confronts climate change: a bottom-up perspective*. New York: Routledge.

Kozubek, Jim. 2016. *Modern prometheus: Editing the human genome with crispr-cas9*. Cambridge: Cambridge University Press.

Krasilovsky, Alexis. 2017. *Great adaptations: Screenwriting and global storytelling*. New York: Routledge.

Kristoff, Nicholas. 1993. China Sees 'Market-Leninism' as Way to Future. *The New York Times*. September 6th, 1993.

Lanouette, William. n.d. *The Scientists' Petition': A forgotten wartime protest.* Retrieved from:http://www.chino.k12.ca.us/cms/lib8/CA01902308/Centricity/Domain/1456/Atomic%20Scientist%20Petition.pdf. Accessed 23 Dec 2013.

Lehmann, Evan & Climatewire. 2010. Who funds contrariness on climate change?. *Scientific American*. http://www.scientificamerican.com/article/who-funds-contrariness-on/

Libby, Leona. 1970. *Fifty environmental problems of timely importance*. Santa Monica: RAND Corporation.

———. 1979. *The uranium people*. New York: Crane Russak & Charles Scribner's Sons.

———. 1983. *Past climates: Tree thermometers, commodities, and people*. Austin: University of Texas Press.

Lilla, Mark. 2016. The end of identity liberalism. *New York Times*. Sunday, November 18th, 2016.

Lincoln, Abraham. 1980. *Address at Gettysburg*. At the sign of the Bud.

Lizza, Ryan. 2009. As the world burns: How the senate and the white house missed their chance to deal with climate change. *The New Yorker*. October 11th, 2010.

Lovelock, James. 1995. *The ages of Gaia: A biography of our living earth*. Oxford: Oxford University Press.

Ludendorff, E. 1936. The nation at war: By general Ludendorff [English version of Der totale Krieg, 1935] Hutchinson & Company, Limited.

MacAskill, Ewen 2008. Hurricane Gustav: Republican convention thrown into Chaos. *The Guardian*, 31 August, 2008. https://www.theguardian.com/world/2008/sep/01/usa.republicans2008

Mackaye, Benton. 1951. Toward Global Law. First published in The Survey, Vol. LXXXVII, No. 6. June, 1951. Republished in *From Geography to Geotechnics* (1968). Champaign: University of Illinois Press.

Mann, Thomas. [1901] 1983. *Buddenbrooks*. Trans. H. T. Lowe-Porter. New York: Alfred A. Knopf.

Mann, Charles C. 2012. The state of the species. *Orion Magazine*. October 24th, 2012. https://orionmagazine.org/article/state-of-the-species/. Accessed 29 Aug 2017.

Margulis, Lynn, and Dorian Sagan. 1995. *What is life?* New York: Simon and Schuster.

Masters, Dexter and Katherine Way, eds. [1946] 2007. *One world or none*. New York: The New Press.

Mastny, Vojtech, and Liqun Zhu. 2014. *The legacy of the cold war: Perspectives on security, cooperation, and conflict*. Plymouth: Lexington Books.

Maxwell, James, and Forrest Bricoe. 1977. There's money in the air: The CFC ban and dupont's regulatory strategy. In *Business Strategy and the Environment*, vol. 6, 276–286.

Maxwell, James and Forrest Briscoe. 1997. "There's money in the air: The CFC Ban and Dupont's regulatory strategy" *Business Strategy and the Environment*, Vol. 6.

Mayer, Milton. 1992. *Robert maynard hutchins: A memoir*. Oakland: University of California Press.

Mazower, Mark. 2009. *No enchanted palace: The end of empire and the ideological origins of the united nations*. Princeton: Princeton University Press.

———. 2012. *Governing the world: The history of an idea*. New York: The Penguin Press.

McGrew, A.G., and P. Lewis. 2013. *Global politics: globalization and the nation-state*. John Wiley & Sons.

McKibben, Bill. 1989. *The end of nature*. New York: Random House.

———. 2010. *Earth: Making life on a tough new planet*. New York: Henry Holt and Company.

McNeill, J. R. 2001. *Something new under the sun: An environmental history of the twentieth-century world (the global century series)*. WW Norton & Company.

McNeill, J.R. 2010. The environment, environmentalism, and international society in the long 1970s. In *The shock of the global. The 1970s in perspective*, ed. Niall Ferguson et al. Cambridge: Harvard University Press.

Meadows, Donella, et al. 1982. *Groping in the dark: The first decade of global modeling*. New York: John Wiley and Sons.

Merton, Robert K. 1976. *Sociological ambivalence and other essays*. New York: Simon and Schuster.

Mirsky, Steven. 2017. Dirty doctors finished what an assassin's bullet started: Disregarding new scientific information can be deadly. *Scientific American*. February 1st, 2017.

Monbiot, George. 2003. *The age of consent*. London: Flamingo.

Monnet, Jean. Richard Mayne, trans. (1978). Memoirs. New York: Doubleday & Company.

Mumford, Lewis. Winter 1964. Authoritarian and democratic technics. Technology and Culture. Vol. 5, No. 1.

———. 1995. A disciple's rebellion. In *Lewis mumford and patrick geddes: The correspondence*, ed. Frank G. Novak Jr., 345. London: Routledge.

———. 2010. *Technics and civilization*, 156. Chicago: University of Chicago Press.

NASA. Earth observing system, mission profiles. http://eospso.gsfc.nasa.gov/eos_homepage/mission_profiles/index.php

Nash, P. 1949. *Outline, an autobiography: And other writings*, 211. London: Faber & Faber.

Nehru, Jawaharlal. 1946. (Reprint 2004) The discovery of India. New Delhi Penguin Books.

Niebuhr, Reinhold. 1949. The illusion of world government. *Foreign Affairs*. April, 1949.

Nolan, Christopher with Lynda Obst and Emma Thomas (Producers) & Nolan, Christopher. 2014. *Interstellar*. United States: Paramount Pictures.

Nussbaum, Martha C. 2010. Kant and Cosmopolitanism. In *The cosmopolitan reader*, ed. David Held and Garrett Wallace Brown. Cambridge: Polity Press.

Oakes, Jason. 2016. Garrett Hardin's tragic sense of life. *Endeavour* 40 (4): 238–247.

Open Secrets. Campaign contribution profile for Senator James Inhofe. https://www.opensecrets.org/members-of-congress/summary?cid=N00005582&cycle=CAREER. Accessed 14 Jan 2018.

Oreskes, Naomi, and Erik M. Conway. 2011. *Merchants of doubt: How a handful of scientists obscured the truth on issues from tobacco smoke to global warming*. New York: Bloomsbury.

Orwell, George. [1945] 1968. You and the atomic bomb" in Sonia Orwell and Ian Angus, eds. *In Front of your nose,* 1945–1950. London: Secker & Warburg.

Ostrom, Elinor. 2009. "*Beyond markets and states: Polycentric governance of complex economic systems*" Nobel Prize Lecture, December 8, 2009 https://www.nobelprize.org/nobel_prizes/economic-sciences/laureates/2009/ostrom_lecture.pdf. Accessed 11 Jan 2018.

Overy, Richard. 2009. *The twilight years: The paradox of britain between the wars*, 93–135. New York: Viking.

Owen, John M. 2010. *Religion, the enlightenment, and the new global order*. New York: Columbia University Press.

Pauling, Linus. December 1963. *Science and peace*. Nobel lecture. http://www.nobelprize.org/nobel_prizes/peace/laureates/1962/pauling-lecture.html

Peters, Adele. 2016. Democracy is getting a reboot on the blockchain. *Fast Company*. https://www.fastcompany.com/3062386/democracy-is-getting-a-reboot-on-the-blockchain.

———. 2017. *The case for eliminating countries and instituting a global democracy*. *Fast Company*. https://www.fastcompany.com/3067153/the-case-for-eliminating-countries-and-instituting-a-global-democracy.

Peterson, T.C., W.M. Connolley, and J. Fleck. 2008. The Myth of the 1970s global cooling scientific consensus. *Amer Meteor Soc* 89: 1325–1338. https://doi.org/10.1175/2008BAMS2370.1.
Pfau, R. 1984. *No sacrifice too great: The life of Lewis L. Strauss*. Charlottesville: University of Virginia Press.
Pinker, Steven. 2018. *Enlightenment now: The case for reason, science, humanism, and progress*. New York: Viking.
Pomeranz, Kenneth, and Steven Topik. 2006. *The world that trade created: society, culture, and the world economy*. London: M. E. Sharpe.
Poole, Robert. 2008. *Earthrise: How man first saw the Earth*. New Haven: Yale University Press.
Popper. 1962. *The open society and its enemies*. Volume One. Princeton University Press, p. 161, 289, 288, 137.
Purdy, Jedediah. 2015. Environmentalism's racist history. *The New Yorker.* August 13th, 2015.
Rayner, T., & Jordan, A. 2016–08-05. *Climate change policy in the European Union*. Oxford Research Encyclopedia of Climate Science. Retrieved 19 Jul. 2017, from http://climatescience.oxfordre.com/view/10.1093/acrefore/9780190228620.001.0001/acrefore-9780190228620-e-47.
Reves, Emery. 1942. *A democratic manifesto*. New York: Random House.
———. 1945. *The anatomy of peace*. New York: Harper and Brothers.
Ricks, Thomas. 2017. *Churchill and orwell: The fight for freedom*. New York: Penguin Press.
Riesbeck, David J. 2016. *Aristotle on political community*. Cambridge: Cambridge University Press.
Robertson, Thomas. 2012. *The malthusian moment: Global population growth and the birth of american environmentalism*. New Brunswick: Rutgers University Press.
Romm, Joseph. 2016. *Climate change: What everyone needs to know*. Oxford: Oxford University Press.
Rosenboim, Or. 2017. *The emergence of globalism: Visions of world order in Britain and the United States, 1939–1950*. Princeton: Princeton University Press.
Roszak, T. 1995. *The making of a counter culture: Reflections on the technocratic society and its youthful opposition*. Berkeley: University of California Press.
Ruddiman, W.F. 2005. How did humans first alter global climate? *Scientific American* 292 (3): 46–53.
———. 2005a. *Plows, plagues, and petroleum: How humans took control of climate*. Princeton: Princeton University Press.
Runciman, David. 2013. *The confidence trap: A history of democracy in Crisis from world war I to the present*. Princeton: Princeton University Press.
Ruse, Michael. 2013. *The gaia hypothesis: Science on a Pagan Planet*. Chicago: University of Chicago Press.
Russell, Bertrand. 1922. *The problem of China*. New York: Century Co.
———. 2001a. *The scientific outlook*, 216. London: Routledge.
Russell, E. 2001b. *War and nature: Fighting humans and insects with chemicals from World War I to Silent Spring*. Cambridge Cambridge University Press(broader transatlantic discourse about total war and the transformation of nature in both popular science and popular culture in the period between 1914 and 1945).
Safire, William. 1982. Come to 'club seabed'. *The New York Times*. November 8th, 1982.
Sanderson, Warren C. 2004. *The end of world population growth in the 21st Century: New challenges for human capital formation and sustainable development*. Routledge. Co.
Schulman, Bruce. 2001. *The seventies: The great shift in American culture, society, and politics*. New York: Simon and Schuster.
Schweickart, Russell L. (2009). Quoted in the film *Earth Days*, directed by Robert Stone. WGBH.
———. (2011). Personal interview via email. December 19th, 2011.
Scott, James Brown. 2000. *Francisco de vittoria and his law of nations*. Union: The Lawbook Exchange, Ltd.
Scripps Institution of Oceanography. 2017. "Keeling Curve" January, 2017. https://scripps.ucsd.edu/programs/keelingcurve/. Accessed 23 Jan 2018.
Shapiro, Judith. 2001. *Mao's war against nature: Politics and the environment in revolutionary China*. Cambridge: Cambridge University Press.

Southern Poverty Law Center. 2017. https://www.splcenter.org/fighting-hate/extremist-files/individual/garrett-hardin Accessed 11 Aug 2017.
Standage, Tom. 1998. *The victorian internet: The remarkable story of the telegraph and the nineteenth century's online pioneers*. New York: Bloomsbury.
Stanley, Alessandra. 1997. From perestroika to pizza. *The New York Times*. December 3rd, 1997. http://www.nytimes.com/1997/12/03/world/from-perestroika-to-pizza-gorbachev-stars-in-tv-ad.html.
Stanton, Elizabeth Cady. [1848] 2015. A declaration of sentiments and resolutions. Carlisle: Applewood Books.
Stella, Tiziana. 2017. Personal interview via email. July 20th, 2017.
Streit, Clarence K. 1939. *Union now: A proposal for a federal union of the democracies of the North Atlantic*. London: Jonathan Cape.
Suskind, R. 2004. Faith, certainty and the presidency of George W. Bush. *New York Times*, p. *17*.
Szilard, L. 1987. *Toward a livable world: Leo Szilard and the crusade for nuclear arms control*. Vol. 3. Cambridge, MA: MIT Press.
Tägil, S. 1998. Alfred Nobel's thoughts about War and Peace. *Nobel prize. org*. *Technological Risk Versus Natural Catastrophe* (RAND, 1971).
Talbott, Strobe. 2009. *The great experiment: The story of ancient empires, modern states, and the quest for a global nation*. New York: Simon & Schuster.
Tennyson, Alfred. 1842. Locksley Hall. *Poems of Tennyson*.
The Social Contract. 1997. Interview with Garrett Hardin. Volume 8, Number 1. Fall 1997. Petoskey, Michigan: The Social Contract Press .
Turner, F. 1893. *The significance of the Frontier in American history*. https://sourcebooks.fordham.edu/mod/1893turner.asp. Accessed 29 May 2017.
VanDevelder, Paul. 2009. *Coyote Warrior*. Nebraska: University of Nebraska Press.
Veblen, T. 1921. *The engineers and the price system*. BW Huebsch, Incorporated.
Vonnegut, Kurt. 1974. *Wampeters, Foma & Granfalloons*. New York: Delacorte Press.
Wagner, Gernot, and Martin L. Weitzman. 2015. *Climate shock: The economic consequences of a Hotter Planet*. Princeton: Princeton University Press.
Ward, Barbara, and René Dubos. 1972. *Only one Earth*. Harmondsworth: Penguin.
Weart, Spencer. 1998. *Never at war: Why democracies will not fight one another*. New Haven/London: Yale University Press.
Weart, Spencer R. 2008. *The discovery of global warming*, 21. Cambridge, MA: Harvard University Press.
Weber, Max. [1919] 2004. *The vocation lectures*. Indianapolis: Hackett Classics.
Wells, H.G. 1914. *The world set free: A story of Mankind*. London: Macmillan & Co.
White, E.B. 1946. *The wild flag*. New York: Houghton Mifflin Company.
———. 1948. Review of *no place to hide*. *The New Yorker*. December 4th, 1948.
Whitman, Walt. 1982. *Walt Whitman: Poetry and prose*. New York: Library of America.
Williams, R. 2013. *The triumph of Human Empire: Verne, Morris, and Stevenson at the end of the world*. Chicago: University of Chicago Press.
Wilson, Woodrow 1918. *President Woodrow Wilson's fourteen points*. The Avalon Project, Yale University. http://avalon.law.yale.edu/20th_century/wilson14.asp
Wilson, G. 2014. *Deepwater horizon and the law of the sea: Was the cure worse than the disease*. *BC Envtl. Aff. L. Rev.*, p. *41*.
Wing-Tsit, Chan. 1963. *A source book in Chinese Philosophy*, 220. Princeton University Press: Princeton.
Wittner, L.S. 1993. *One world or none: A history of the world nuclear disarmament movement through 1953*. Palo Alto: Stanford University Press.
World Health Organization. 2005. *Chernobyl: The true scale of the accident*. Geneva: Joint News Release WHO/IAEA/UNDP. September 5th, 2005. *http://www.who.int/mediacentre/news/releases/2005/pr38/en/*
Wright, Robert. 2001. *Nonzero: The logic of human destiny*. New York: Vintage Books.
Your Country's Call. 1915. Imperial war museum. http://www.iwm.org.uk/collections/item/object/27751. Accessed 29 May 2017.

# Index




R. S. Deese, *Climate Change and the Future of Democracy*, Environmental Challenges and Solutions 5, https://doi.org/10.1007/978-3-319-98307-3